ACCLAIM FOR JEFF GOODELL
AND **BIG COAL**

"A fascinating and frightening glimpse into the hidden power circuits of American industry and politics."
— JAMES HOWARD KUNSTLER, author of *The Long Emergency*

"Smart, fair, impassioned, and very well written, this is a book that matters." — MARK HERTSGAARD, author of *Earth Odyssey*

"The book's strength lies in Goodell's ability to connect our mundane daily activities, such as flipping on the living room lights and powering up our laptops, with the grimy business that powers these things."
— *Washington Post Book World*

"A compelling and frank analysis of the 'black rock.'" — *Seed*

"There is perhaps no greater act of denial in modern life than sticking a plug into an electrical outlet . . . Jeff Goodell breaks the spell with a single number: 20. That's how many pounds of coal each person in the United States consumes, on average, every day to keep the electricity flowing. . . . Goodell's writing [is] so fiery and committed."
— *New York Times Book Review*

"An absorbing, urgently important book. Jeff Goodell does a marvelous job exposing the hidden workings of a deeply entrenched industry and of showing how our use of coal poses a grave threat to our collective future."
— BARBARA FREESE, author of *Coal: A Human History*

"Well researched and very well written." — *Pittsburgh Post-Gazette*

"Goodell muckrakes in the tradition of Ida Tarbell, Rachel Carson, and Eric Schlosser, leading us confidently, if ruefully, on a tour through the world of coal." — *Plenty*

"Persuasive journalism at its best . . . Goodell succeeds by seamlessly weaving together the political and economic issues and the health and environmental costs of coal." — *Columbia Journalism Review*

"Jeff Goodell's incisive, gripping first-hand report on the second coming of King Coal impacts everyone and everything on earth." — RALPH NADER

"[Goodell] tracks the black rock on a bracing, eye-opening journey — from deep underground, through mining towns, to the railroads, through the halls of Congress, and, finally, into the air." — *Audubon*

"The perfect book with which to develop the necessary, clear understanding of the environmental and social consequences that accompany a seemingly simple gesture like plugging an appliance into the wall."
— *Brooklyn Rail*

"*Big Coal*'s greatest strength lies in Goodell's ability to tell human stories — how individuals, families, and communities are affected by the mining, production, and consumption of coal." — *Rocky Mountain News*

"Enlightening and disturbing." — *Rolling Stone*

"We might think of coal as old-fashioned, but Goodell reveals that coal — yes, coal — is being revived as an energy source and warns of the danger that dependence on those dusty little black clumps poses to the earth's future." — *Associated Press*

"The hidden truth about coal — the dangers to miners, health risks from air pollution and accumulating greenhouse gasses — is what Jeff Goodell is after in his groundbreaking book." — *BookPage*

"What keeps the book going is its slangy voice, full of well-reasoned outrage and a deep sense of what's at stake." — *Orion*

"Goodell does a first-rate job of balancing environmental concerns with interviews from the human faces associated with Big Coal ... The book opens our eyes to how we can improve our use of coal and figure out other, less destructive ways to create the energy we need."
— *Library Journal*

"Compelling." — *Publishers Weekly*, starred review

"Eye-opening and provocative ... Goodell is right to say that the coal economy is little documented and not well understood, but his book makes a welcome corrective." — *Kirkus Reviews*

BIG COAL

BOOKS BY JEFF GOODELL

THE CYBERTHIEF AND THE SAMURAI

SUNNYVALE: THE RISE AND FALL OF A
SILICON VALLEY FAMILY

OUR STORY: SEVENTY-SEVEN HOURS THAT
TESTED OUR FRIENDSHIP AND OUR FAITH

BIG COAL: THE DIRTY SECRET BEHIND
AMERICA'S ENERGY FUTURE

BIG COAL

THE DIRTY SECRET
BEHIND AMERICA'S
ENERGY FUTURE

JEFF GOODELL

A MARINER BOOK
HOUGHTON MIFFLIN COMPANY
Boston ▪ *New York*

FOR GRACE

First Mariner Books edition 2007
Copyright © 2006 by Jeff Goodell
ALL RIGHTS RESERVED

For information about permission to reproduce selections from
this book, write to Permissions, Houghton Mifflin Company,
215 Park Avenue South, New York, New York 10003.

Visit our Web site: www.houghtonmifflinbooks.com.

Library of Congress Cataloging-in-Publication Data
Goodell, Jeff.
Big coal : the dirty secret behind America's
energy future / Jeff Goodell.
p.   cm.
Includes bibliographical references and index.
ISBN-13: 978-0-618-31940-4
ISBN-10: 0-618-31940-9
1. Coal mines and mining — United States.
2. Coal — United States. 3. Energy minerals —
United States. I. Title.
TN805.A5G665 2005
333.793'20973 — dc22   2005033199

ISBN-13: 978-0-618-87224-4 (pbk.)
ISBN-10: 0-618-87224-8 (pbk.)

PRINTED IN THE UNITED STATES OF AMERICA

Book design by Robert Overholtzer

DOC 10 9 8 7 6

ILLUSTRATION CREDITS: page 1, photo reprinted courtesy of
Vivian Stockman/www.ohvec.org; page 7, map reprinted courtesy of the
U.S. Geological Survey; page 95, photo reprinted courtesy of Steve Roe;
page 98, graphic reprinted courtesy of TVA; page 171, photo
reprinted courtesy of Tracy Chu.

America's epic is the odyssey of appetite.

—*Campbell McGrath, "Infinite Needs"*

Contents

Introduction

ONE OF THE TRIUMPHS of modern life is our ability to distance ourselves from the simple facts of our own existence. We love our hamburgers, but we've never seen the inside of a slaughterhouse. We're not sure if the asparagus that accompanies our salmon is grown in Ecuador or Oregon. We flush the toilet and don't want to know any more. If we feel bad, we take a pill. We don't even bury our own dead — they are carted away and buried or burned for us.

It's easy to forget what a luxury this is — until you visit a place like China. Despite its booming economy in recent years, the insulating walls of modern life have not yet been fully erected there. In restaurants, the entrées are often alive in a cage in the dining room. Herbs and acupuncture needles inspire more faith than pharmaceutical drugs. Toilets stink. In rural areas, running water is a surprise, hot water a thrill. When you flip the switch on the wall and the light goes on, you know exactly what it costs — all you have to do is take a deep breath and feel the burn of coal smoke in your lungs.

To a westerner, nothing is more uncivilized than the sulfury smell of coal. You can't take a whiff without thinking of labor battles and underground mine explosions, of chugging smokestacks and black lung.

But coal is everywhere in twenty-first-century China. It's piled up on sidewalks, pressed into bricks and stacked near the back doors of homes, stockpiled into small mountains in the middle of open fields, and carted around behind bicycles and old wheezing locomotives. Plumes of coal smoke rise from rusty stacks on every urban horizon. There is soot on every windowsill and around the collar of every white shirt. Coal is what's fueling China's economic boom, and nobody makes any pretense that it isn't. And as it did in America one hundred years ago, the power of coal will lift China into a better world. It will make the country richer, more civilized, and more remote from the hard facts of life, just like us.

The cost of the rough journey China is undertaking is obvious. More than six thousand workers a year are killed in China's coal mines. The World Health Organization estimates that in East Asia, a region made up predominantly of China and South Korea, 355,000 people a year die from the effects of urban outdoor air pollution. The first time I visited Jiamusi, a city in China's industrial north, it was so befouled by coal smoke that I could hardly see across the street. All over China, limestone buildings are dissolving in the acidic air. In Beijing, the ancient outdoor statuary at a 700-year-old Taoist temple I visited was encased in Plexiglas to protect it. And it's not just the Chinese who are paying for their coal-fired prosperity. Pollution from China's power plants blows across the Pacific and is inhaled by sunbathers on Malibu beach. Toxic mercury from Chinese coal finds its way into polar bears in the Arctic. Most seriously, the carbon dioxide released by China's mad burning of coal is helping to destabilize the climate of the entire planet.

All this would be much easier to condemn if the West had not done exactly the same thing during its headlong rush to become rich and prosperous. In fact, we're still doing it. Although America is a vastly richer country with many more options available to us, our per capita consumption of coal is three times higher than China's. You can argue that we manage it better — our mines are safer, our power plants are cleaner — but mostly we just hide it better. We hide it so well, in fact, that many Americans think that coal went out with corsets and top hats. Most of us have no idea how central coal is to our everyday lives or what our relationship with this black rock really costs us.

In truth, the United States is more dependent on coal today than ever before. The average American consumes about twenty pounds of it a day. We don't use it to warm our hearths anymore, but we burn it by wire whenever we flip on the light switch or charge up our laptops. More than one hundred years after Thomas Edison connected the first light bulb to a coal-fired generator, coal remains the bedrock of the electric power industry in America. About half the electricity we consume comes from coal — we burn more than a billion tons of it a year, usually in big, aging power plants that churn out amazing quantities of power, profit, and pollution. In fact, electric power generation is one of the largest and most capital-intensive industries in the country, with revenues of more than $380 billion in 2005. And the rise of the Internet — a global network of electrons — has only increased the industry's power and influence. We may not like to admit it, but our shiny white iPod economy is propped up by dirty black rocks.

This was not how things were supposed to go in America. Coal was supposed to be the engine of the industrial revolution, not the Internet revolution. It once powered our steamships and trains; it forged the steel that won the wars and shaped our cars and skyscrapers and airplanes. It kept pioneers warm on the prairie and built fortunes for robber barons such as Henry Frick and Andrew Carnegie. Without coal, the world as we know it today would be impossible to imagine. There should be monuments to coal in every big city, giant statues of Pennsylvania anthracite and West Virginia bituminous. It is literally the rock that built America.

But we've been hooked on coal for almost 150 years now, and like a Bowery junkie, we keep telling ourselves it's time to come clean, without ever actually doing it. We stopped burning coal in our homes in the 1930s, in locomotives in the 1940s, and by the 1950s it seemed that coal was on its way out for electricity generation, too. Nuclear power was the great dream of the post–World War II era, but the near-meltdown of the Three Mile Island nuclear plant in 1979 put an end to that. Then natural gas overtook coal as the fuel of choice. If coal was our industrial smack, natural gas was our methadone: it was clean, easy to transport, and nearly as cheap as coal. Virtually every power plant built in America between 1975 and 2002 was gas-fired. Almost everybody in the energy world presumed that the natural gas era would soon give way to even cleaner sources of power generation — wind, solar, biofuels, hydrogen, perhaps someday solar panels on the moon. As for the old coal plants, they would be dismantled, repowered, or left to rust in the fields.

But like many revolutions, this one hasn't progressed quite as planned.

Energy-wise, the fundamental problem in the world today is that the earth's reserves of fossil fuels are finite but our appetite for them is not. The issue is not simply that there are more people in the world, consuming more fossil fuels, but that as economies grow and people in developing nations are lifted out of poverty, they buy cars and refrigerators and develop an appetite for gas, oil, and coal. Between 1950 and 2000, as the world population grew by roughly 140 percent, fossil fuel consumption increased by almost 400 percent. By 2030, the world's demand for energy is projected to more than double, with most of that energy coming from fossil fuels.

Of course, every barrel of oil we pump out of the ground, every cubic foot of natural gas we consume, and every ton of coal we burn further depletes reserves. For a while, our day of reckoning was put off by the fact that technological innovation outpaced consumption: the more fossil fuels we burned, the better we became at finding more, lulling us into a false belief that the world's reserves of fossil fuels are eternal. But that delusion can't last forever. In fact, there are increasing signs that it won't last much longer.

Oil is the most critical fossil fuel for modern economies, underlying everything from transportation to manufacturing. In 2005, the world consumed more than 82 million barrels of oil each day, about 30 percent of which came from the Middle East. The world is not going to run out of oil anytime soon, but it might run out of cheap, easy-to-get oil. As that happens, prices are likely to spike, fundamentally disrupting major parts of the world's economy. You don't have to buy into the apocalyptic scenarios that some doomsayers predict — the collapse of industrial society, widespread famine — to see that the end of cheap oil is going to inspire panic and economic chaos as the world scrambles to find a replacement energy source.

The situation with natural gas is not much better. In the United States, consumption of natural gas, which is mostly used for home heating and the manufacture of industrial products, as well as agricultural fertilizers and chemicals, has jumped by about 40 percent in the past two decades. About 85 percent of that gas came from domestic sources, but production in the United States has been flat for several decades, leading us to import more and more from Canada, where production is also beginning to peak. There are still substantial reserves in places such as Russia and Qatar, but the global shipping and trading infrastructure is woefully undeveloped. Upgrading it will cost billions of dollars and take decades to complete. Not surprisingly, natural gas prices have tripled in the past few years and caused home heating bills to rise rapidly in many regions of the country.

What about the other alternatives? Nuclear power can be used to generate electricity, but no new plants have been built in America in thirty years. This is primarily because nuclear plants are still haunted by the ghosts of Three Mile Island and Chernobyl, as well as unresolved problems of radioactive waste. Even if the social and environmental hurdles could be overcome, nuclear plants are so expensive to build that a major resurgence is unlikely. And as much as we would all like to imagine we could live in a world powered by solar panels and wind turbines, these alternative energy sources are not yet capable of powering our high-tech economy.

Out of this, coal has emerged as the default fuel of choice. Coal has a number of virtues as a fuel: it can be shipped via boats and railroads, it's easy to store, and it's easy to burn. But coal's main advantage over other fuels is that it's cheap and plentiful. There are an estimated 1 trillion tons of recoverable coal in the world, by far the largest reserve of fossil fuel left on the planet. And despite a run-up in coal prices in 2004 and 2005, coal is still inexpensive compared to other fuels. In a world starved for energy, the importance of this simple fact cannot be underestimated: the world needs cheap power, and coal can provide it.

America is literally built upon thick seams of coal. Just as Saudi Arabia dominates the global oil market because of the geological good luck of having more than 20 percent of the world's oil reserves, the

United States is a big advocate for coal because it has the geological good luck of having more than 25 percent of the world's recoverable coal reserves — about 270 billion tons — buried within its borders. As coal industry executives never tire of pointing out, this is enough coal to fuel America at the current rate of consumption for about 250 years. To put the size of its bounty into perspective, consider this: all of western Europe has only 36 billion tons of recoverable coal. China has less than half as much as the United States — 126 billion tons. India and Australia, both big coal burners, have even less than China. The only country with reserves that come close to America's is Russia, with 176 billion tons, but much of that coal is in remote regions and difficult to mine. Not surprisingly, coal boosters often refer to America as "the Saudi Arabia of coal."

America's great bounty of coal confers upon the United States many economic and political advantages. As a purely practical matter, it means that America will not go dark while scientists search for a replacement for fossil fuels. If the world becomes energy-starved, our reserves mean that America will have a source of fuel to keep our factories running and our cities well lit. If oil supplies collapse and prices skyrocket, we can begin a crash program to build coal liquefaction plants, which can turn coal into synthetic diesel. It won't keep our SUVs rolling, but it might help keep our F-16s flying. Using a similar process, coal can also be transformed into synthetic natural gas, fertilizers, and a variety of industrial chemicals.

But this great bounty of coal is also a great liability. It means that America has a big incentive to drag out the inevitable transition to cleaner, more modern forms of energy generation. In a world that is moving toward energy efficiency, coal is a big loser. Alternative energy guru Amory Lovins estimates that by the time you mine the coal, haul it to the power plant, burn it, and then send the electricity out over the wires to the incandescent bulb in your home, only about 3 percent of the energy contained in a ton of coal is transformed into light. In fact, just the energy wasted by coal plants in America would be enough to power the entire Japanese economy. In effect, America's vast reserve of coal is like a giant carbon anchor slowing down the nation's transition to new sources of energy. And because coal is the dirtiest and most carbon-intensive of all fossil fuels — coal plants are responsible for nearly 40 percent of U.S. emissions of carbon dioxide, the main greenhouse gas — a commitment to coal is tantamount to a denial of a whole host of environmental and public health issues, including global warming. When you're sitting on top of 250 years' worth of coal, an international agreement to limit carbon dioxide emissions, such as the Kyoto Protocol, is easily seen as a crude attempt by jealous competitors to blunt one of America's great strategic and economic advantages.

In America, the story of coal's emergence as the default fuel of choice is inextricably tied up with corruption, politics, and war. California's long, torturous "energy crisis," which lasted through the summer of 2000 and culminated in rolling blackouts in January 2001, underscored the need for new investment in electricity generation and transmission. The collapse of Big Coal's arch-nemesis, Enron, also helped coal regain some of its luster. Once heralded as a great modernizing force in the electric power industry, promising to bring a market-driven revolution to the old energy empire, Enron turned out to be a den of thieves. The company's fall — one of the largest bankruptcies in U.S. history — helped throw the natural gas market into turmoil, sending prices skyrocketing and making coal so inexpensive in comparison that operating a coal plant became, as one industry consultant explained it to me, "like running a legal mint."

The 2000 presidential election was another turning point. Democratic candidate Al Gore was one of the first American politicians to take global warming seriously, and anyone who takes global warming seriously is not a friend of Big Coal. Coal industry executives knew that if Gore was elected, regulations to limit or tax carbon dioxide emissions wouldn't be far behind. So Big Coal threw its money and muscle behind George W. Bush, helping him gain a decisive edge in key industrial states, including West Virginia, a Democratic stronghold that had not voted for a Republican presidential candidate in seventy-five years. After the disputed Florida recount, West Virginia's five electoral votes provided the margin that Bush needed to take his seat in the Oval Office.

President Bush made good on his debt. Within weeks of taking the oath of office, Bush began staffing regulatory agencies with for-

THE POLITICS OF BIG COAL

Federal Campaign Contributions to Parties and Candidates 2000–2004

In Total Dollars

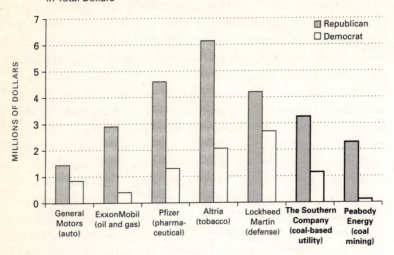

....and by Percentage of Company Profits

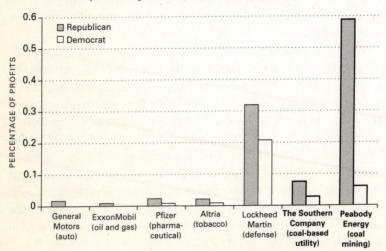

TOP CONTRIBUTORS IN MAJOR U.S. INDUSTRIES

Source: Center for Responsive Politics, EDGAR

mer coal industry executives and lobbyists. Not surprisingly, Big Coal also played a prominent role in Vice President Dick Cheney's National Energy Policy Development Group, which was charged with crafting a new energy policy. The task force's recommendations were unabashedly coal-friendly, including a call for up to 1,900 new power plants over the next twenty years; a $2 billion, ten-year subsidy for "clean coal" technology; and a recommendation that the Department of Justice "review" enforcement actions against dirty coal burners.

Finally, the terrorist attack on the World Trade Center on September 11, 2001, was an unexpected boon for Big Coal. Politically, it took the spotlight off many of the Bush administration's controversial coal-friendly energy policies, which were just beginning to make headlines. More important, 9/11 changed the tone of the debate about energy in America, making many of us reconsider the high cost of our dependence on oil from the Middle East. In our globally connected world, "energy independence" is more of a political slogan than a practical reality. But as long as American soldiers were dying in the oil-rich Middle East, it seemed downright unpatriotic to oppose coal.

For Big Coal, this change in America's political and economic climate was transformative. Around the country, any open patch of ground near a railroad, a high-voltage transmission line, and a decent-size population of electricity consumers became a possible site for a new coal plant. As of 2006, more than 150 new plants, representing more than $130 billion in new investment, were either planned or under construction in the United States. Long-shuttered mines were reopening, and old coal miners were dusting off their boots. Wall Street analysts, in a swoon over the old rock (the Street loves big, expensive projects with all-but-guaranteed returns such as coal plants), began cranking out pro-coal reports with titles such as "Come On Over to the Dark Side" and "Party On, King Coal!" The rebirth of coal is not just about energy; it is also a cultural uprising of sorts, a taking back of a key part of America's economic life that is, in its own way, as reactionary as the public campaigns against evolution or gay marriage. It is about the revenge of the Old Economy over all those technology-loving geeks who thought an energy revolution was at hand, who said that the forces of creative destruction would wreak havoc on one of the world's great industrial empires, and who naively believed that the future would be powered by solar panels and bio-diesel.

Lost in the hype, of course, is a sober accounting of what this new coal boom might really cost us. In January 2006, seventeen men died in Appalachian coal mines, including twelve men in an explosion at the Sago mine in northern West Virginia and two more after a fire in the Alma mine in the southern part of the state. Since 1900, more than 100,000 people have been killed in coal mine accidents, many forever entombed by collapsed roofs and tumbling pillars. Black lung, a disease common among miners from inhaling coal dust, can be conservatively estimated to have killed another 200,000 workers. And burning coal is even more deadly. In just the past twenty years, air pollution from coal plants has shortened the lives of more than half a million Americans. The broad legacy of environmental devastation — acid rain, polluted lakes and rivers, mined-out mountains — is impossible to tabulate. In Appalachia alone, the waste from mountaintop removal mining (instead of removing the coal from the mountain, the mountain is removed from the coal) has buried more than 1,200 miles of streams, polluted the region's groundwater and rivers, and turned about 400,000 acres of some of the world's most biologically rich temperate forests into flat, barren wastelands. Plumes of toxic particles drift from Ohio northeast to Maine; a molecule of mercury emitted from the stack of a power plant in Tampa ends up in the brain of a child in Minneapolis. If and when fruit trees start growing on the Alaskan tundra, American coal burners past and present will be largely responsible.

Not so long ago, you could justify coal's dark side with a single word: jobs. In the 1920s, when more than 700,000 workers worked in the mines, it was plausible to argue that miners were the backbone of the economy. Today there are more florists in America than there are coal miners. And if coal mining were the sure-fire ticket to wealth and prosperity that many in the industry claim, West Virginians would be dancing on gold-paved streets. Over the past 150 years or so, more than 13 billion tons of coal have been carted out of the Mountain State. What do West Virginians have to show for it? The lowest median household income in the nation, a literacy rate in the southern coalfields that's about the same as Kabul's, and a generation of young people who are abandoning their home state to seek their fortunes elsewhere.

The argument that cheap power is vital to keeping American manufacturers competitive also is suspect. At a time when U.S. auto manufacturers spend more money on health care for their workers than on steel for their cars, it's increasingly hard to make the case that cheap electricity is a major factor in keeping jobs from being exported to Asia. By contrast, a full-blown push for clean energy could unleash a jobs bonanza that would make what happened in Silicon Valley in the 1990s look like a bake sale.

What's most remarkable about the coal boom is that, unlike other recent booms, which were driven by an overwhelming exuberance, this one is driven by overpowering fear: fear that the world is running out of energy, fear that America is losing its edge, fear of relinquishing the industrial age belief that we can drill and mine our way to peace and prosperity, and, most of all, fear that if we don't burn more coal, we will put not only the economic health of the nation at risk but civilization itself. "Have you ever been in a blackout?" one coal executive asked me while I was researching this book. "Do you remember how dark the whole world gets? Do you remember how scary it is?"

Growing up in California, I had a firsthand look at the devil's bargain of progress. In the space of a few decades, my hometown of Silicon Valley went from a sleepy oasis of fruit trees to the epicenter of the digital world. The lovely apricot orchards in my neighborhood were bulldozed and replaced by tract housing. Ferraris appeared at stoplights like exotic birds. I saw some friends and family members catch the wave and get rich, while others who had less talent for life in the new world fell farther and farther behind. I loved my computer, and I loved the freedom and prosperity that came with it, but I could never rid myself of the sense that the wonders of the digital world had come at a high cost.

When I began research on this book, I felt an immediate and unexpected connection with many people who had grown up in Appalachian coal towns. Many of them had fled the world they grew up in (as I had) and looked back on it with a particular kind of sorrow that was very familiar to me. This note from Jennifer Stock, a thirty-five-year-old West Virginia native who now lives in Seattle, is typical:

> I grew up in Logan, West Virginia. When I was a teenager, I would go up on Blair Mountain to party. There was a tall fire lookout tower on the top that was great fun to climb. You could see so many ridges from it; the hills just went on and on. Last time I tried to go back there, a few years ago, there were all sorts of fences in the way. The coal companies are as ruthless to the environment now as they used to be toward their "employees." Strip mining and "mountain top removal" are turning the area into a landscape from hell, and to add insult to injury, the profits reaped from these efforts still completely elude the inhabitants of the land. And then these people are blamed for their ignorance and poverty because it is easier for their fellow citizens to think that the ugliness is due to individual moral failing ("lazy rednecks") than [to] the economic system in which we all participate, by which we are all culpable.

Until I was forty years old, I had never seen a lump of coal. As a kid, I'd visited hydroelectric dams in the Sierra Nevada foothills and wind farms above San Francisco Bay. These sights made generating electricity seem easy and natural, like growing wheat or getting a suntan. It gave me the idea — one that I clung to for years — that it really didn't matter if I left the light on in the other room, because it just meant the water turbines and the windmills had to spin a little longer. Of course, this is precisely the kind of blue-state ignorance that red staters scorn, and justifiably so, since the red states often bear the burden of the blue states' cluelessness. (Half the electricity in Los Angeles, for example, is imported from coal-fired power plants in Nevada and New Mexico.) But it is also the kind of cluelessness that power companies have spent years encouraging. If you doubt this, just try deciphering the spinning wheels on the electric meter outside your house. Power companies figured out long ago that the more they isolate consumers from the true costs and consequences of their kilowatts, the more successful the companies will be.

I lost my innocence in the summer of 2001, when the *New York Times Magazine* sent me down to West Virginia to write about the surprising comeback of coal during the early days of the Bush administration. I began my research by visiting one of the largest mines in West Virginia, Hobet 21, which is owned by Arch Coal, the second-biggest coal company in America. When I pulled up to the mine gate, I was a few minutes early for my meeting with a mine engineer, so I got out of the car and wandered around. Down by the railroad tracks, I confronted a large pile of the most beautiful black rocks I had ever seen. They were black beyond black and seemed to pull the light out of the sky around them. It took me a moment to realize that these rocks were coal.

Over the next several weeks, I visited several coal mines and talked with the engineers who worked in them. I drove to Cabin Creek, a narrow valley south of Charleston, West Virginia, where, in 1913, mining company thugs opened fire with Gatling guns on their own workers. I flew in a small plane over the southern coalfields, getting a bird's-eye view of the devastation wrought by mountain-top removal mining. I visited filled-in creeks and drove around with a local politician who explained to me with a straight face that flattening West Virginia was actually a good thing, because the state needed more level ground for golf courses.

All of this was quite eye-opening to me. I felt as if I had stumbled into the gritty underbelly of modern life, the dark, dirty place where the real work is done and the real deals are cut.

The most memorable moment of that trip — and, in some ways, the real beginning of this book — was a dinner I had with Bill Raney, the head of the powerful West Virginia Coal Association. We met at the bar at the Marriott hotel in Charleston, not far from Raney's office. Raney is a short, dapper man with a folksy West Virginia drawl. He was dressed that night in an expensive suit and nice tie, looking more like a Beltway politician than a man who grew up in a coal camp. It was less than a year after the 2000 election, and Raney's Beltway credentials were at an all-time high after his having helped deliver the state of West Virginia — and the Oval Office — to President Bush.

But it wasn't Raney's political connections that impressed me.

Nor was it his defense of mountaintop removal mining as a necessary evil if West Virginia is to compete with coal mines in other states. It was what he said about technology. "The thing that people don't realize," Raney drawled, "is that if it weren't for coal, there would be no Internet, no Microsoft, no Yahoo!" He leaned over his dinner plate. "Did you know that it takes more electricity to charge up a Palm than it does to run an ordinary refrigerator? And that every time you order a book from Amazon, you burn over three pounds of coal?"

I didn't know that. Later, I would find out that his calculations were wildly exaggerated. But his larger point about the interconnectedness of the dirty life of the mines and the sparkly pixels on my computer screen was correct. What Raney was really saying to me, I understood later, was this: You use a computer. You have lights in your house. You watch TV. You are implicated in all of this.

We all are.

I spent three years researching and writing this book. I visited coal mines and power plants in ten states, as well as in China. I rode coal trains across the Great Plains, detonated 55,000 pounds of explosives in Wyoming, and spent a month on a research vessel in the North Atlantic with scientists who were studying climate change. As it turned out, the three years I spent on this book were three of the four hottest years on record. When I started my research, an energy industry consultant joked with me that a ferocious hurricane would have to wipe out New Orleans before America would wake up to the dangers of global warming. By the time I finished the book, that hurricane had arrived, although the awakening had not.

During those three years, about 3 billion tons of coal went up in smoke in America. They created light and heat for much of the nation (not to mention the glow on my computer screen even now as I write). But during those years, the American Lung Association calculates, about 72,000 people in the United States died prematurely from the effects of coal-fired power plant pollution — more than from AIDS, murder, or drug overdose.

Obviously, there's no free power lunch: nukes can melt down, dams flood valleys, and wind turbines kill birds. Building the modern world is fraught with tradeoffs. But unlike in China or India, it's hard to argue that by burning coal to create electricity, America is lifting millions out of poverty and introducing them to hot showers and cold Cokes. Our affection for coal is essentially an old habit and an indulgence. At best, it's a short-term solution to a long-term problem. And the price of this indulgence may be higher than any economist can calculate. Wally Broecker, the great climatologist at Columbia University's Lamont-Doherty Earth Observatory, has compared the earth's climate to a dragon: when you poke it, you can never be sure how it's going to react. As Broecker says, "We're playing with the whole planet, dammit, just to get energy for a few hundred years."

Working on this book, I came to understand that when we talk about energy, what we are really talking about is how we live and what we value. Are we willing to put the earth's climate at risk to save ten bucks on our utility bills? To what degree do we want energy corporations to control our access to power? Is it more important to protect yesterday's jobs or to create a new industry for tomorrow? What degree of sacrifice are we willing to make in our lifestyles to ensure the well-being of our children and grandchildren? The coal industry, of course, would rather keep the conversation focused on the price of electricity per kilowatt-hour and the stockholders' return on investment. Coal is a commodity business, after all, one that is run mostly by number crunchers who see the world as a spreadsheet to conquer. Questions about the price of progress, and how we draw the line between what is acceptable to us as a rich, modern society and what is not, do not fit easily into these calculations.

This problem is as old as our love affair with coal. In 1893, the Chicago World's Fair gave many Americans their first view of the miraculous dynamos that turned coal into clean, bright incandescent light. Among the fair's 23 million visitors was Henry Adams, a well-traveled writer and historian from a prominent Massachusetts family. (His grandfather and great-grandfather were both presidents, and his brother ran the Union Pacific railroad.) For Adams, the sight of the coal-fired dynamos was a sign that American life was about to change irrevocably. He felt the forty-foot dynamos as a moral force, much as the early Christians felt the Cross. And it frightened him. As Adams put it, "Chicago asked in 1893 for the first time the question whether the American people knew where they were driving."

A few words about the organization of this book. I've structured it in three sections that roughly track the life cycle of coal. The first, called "The Dig," deals with the mining and transportation of coal. The second, titled "The Burn," is about the politics of coal-burning power plants and the health effects of air pollution. The final section, called "The Heat," is about coal's role in climate change and how the industry intends to meet (or not meet) this formidable challenge. By organizing the book this way, I hope to give a sense of the broad impact that coal has on our lives. Too often, debates about energy degenerate into arcane discussions about the regulatory minutiae of sulfur dioxide emissions or flaws in the mathematical algorithms used to calculate changes in the earth's average temperature over the past millennium. But coal is not just a form of energy subject to scientific measurement. It is a hidden world unto itself — a world with its own economy, subcultures, and values, yet one whose influence can be felt in every aspect of our lives.

Like every writer, I bring my own baggage to this book. For the record, I am not a member of any environmental organization and never have been. My biases are less political than entrepreneurial. The Silicon Valley town I grew up in may have been full of greedy strivers, but you can't say they lacked vision or a willingness to tackle tough problems. Writing this book, I found myself exploring a world that is the inverse of my hometown, a place where instead the goal often seems to be to explain why a problem can't be solved, or why it's too expensive to solve, or to spin problems into nonproblems. I don't mean to suggest that there aren't lots of well-meaning people in the coal industry or that many of the engineers I met aren't brilliant. Keeping the lights on in a nation of 300 million people is a job that's as challenging and complex in its own way as putting a man on the moon. I mean simply that from the industry's point of view, the goal of technological change is never to reinvent the wheel; it is to figure out new ways to keep the old wheels rolling. This is hardly surprising — what industry plots its own obsolescence? But for me, experiencing the coal industry was a bizarre inversion of the can-do optimism I'd grown up with. I sometimes felt I had stumbled upon a group of mad scientists frantically scheming to invent their own industrial fountain of youth.

Throughout this book, I frequently use the phrase "Big Coal" as shorthand for the alliance of coal mining companies, coal-burning utilities, railroads, lobbying groups, and industry supporters that make the coal industry such a political force in America. The phrase is not meant to suggest that the industry is monolithic, or that they all meet together in smoke-filled rooms to cut deals and hammer out grand strategies. Obviously, there are diverse players in the industry, with diverse points of view. You will meet many of them in this book. But it is also true that the coal industry, like the auto industry, the oil industry, the telecommunications industry, and just about every other multibillion-dollar industry, can be identified by certain common goals and pursuits. The phrase "Big Coal" is meant to suggest that commonality, as well as to remind the reader of the power and influence of the players who are involved.

Finally, a word about the many coal miners, power plant engineers, and railroad workers I met in the course of reporting this book. Whatever criticisms I may have of Big Coal, none of it should be taken as a sign of disrespect for the difficult, dangerous work done by these men and women on the frontlines. Keeping America powered up is not an easy job, and the people who do it deserve our admiration and our thanks. They certainly have mine.

THE DIG

Chapter 1

The Saudi Arabia of Coal

IF YOU WANT TO FEEL the spirit and exuberance of a place that's rich in fossil fuels, you don't have to travel to Dubai. Just visit Gillette, Wyoming, the self-proclaimed Energy Capital of the World. Out here on the vast, open prairie, where the great Sioux warrior Crazy Horse once roamed and not far from where Lieutenant Colonel George Armstrong Custer made his last stand, you don't find many people worrying about the fact that the world may be running out of fossil fuels. Gillette is in the heart of the Powder River basin, a 250-mile-long wind-sculpted hollow between the Bighorn Mountains and Devils Tower National Monument which contains some of the richest reserves of oil, gas, and especially coal that remain in America.

Not surprisingly, living in close proximity to this buried treasure makes everyone into an energy optimist. At a coffee shop in town, I met a woman who told me that when she runs on the treadmill in her living room, she likes to turn all the lights on in the house, as well as the stereo, the TV, and every electronic appliance. "I like to feel the energy," she told me. "It keeps me going." At Energy Dodge, one of several national park–size car dealerships in Gillette, you'll find row after row of 4x4s with big, thirsty Magnum V-8 engines. "You mean 'high-birds'?" a salesman joked when I stopped in to ask if any customers were asking for more fuel-efficient vehicles. "We shoot them out here."

Beneath all the excitement, however, lurks a question: exactly how much oil, gas, and coal is left out there beneath the prairie? The question is especially relevant to coal, which is by far the most bountiful of the three fuels in the region. The best of the oil and gas reserves have already been pumped out in Wyoming, and although there's a lot of oil and gas left, much of it is in restricted areas or hard to get at. But coal! Everyone knows there's a lot of coal in the Powder River basin. All you have to do is drive a mile or so outside town to a viewing platform built on the side of the road. It's the same kind of viewing platform you see in other regions of the country for scenic vistas of spectacular mountains and valleys. In Gillette, the scenic vista is of sixty-foot-thick coal seams in an operating mine.

When you see these great black seams and then imagine them running beneath your feet and under the city of Gillette — and under the sagebrush all the way out to the Bighorn Mountains nearly fifty miles to the west, then another fifty miles to the north beyond the Montana border, and then seventy-five miles or so to the south, nearly down to where the wagon wheels of migrants on the Oregon Trail cut ruts in the rocks — it's easy to see why people who live in the Powder River basin feel so optimistic about coal's place in America's energy future. As Fenimore Chatterton, Wyoming's secretary of state, put it back in 1902, "Coal? Wyoming has enough to run the forges of Vulcan, weld every tie that binds, drive every wheel, change the north pole into a tropical region, or smelt all hell!" A more sober assessment, provided by the U.S. Department of Energy, suggests that Wyoming has about 42 billion tons of mineable coal left in the ground. At the current rate of consumption, that's enough to supply the entire United States for more than forty years.

But it's coming out quickly. Today about 40 percent of the coal burned in America — about 400 million tons — is mined in the Powder River basin. The vast majority of that coal comes from thirteen huge strip mines near Gillette. One of the biggest is Cordero Rojo, which is owned by Kennecott Energy, a subsidiary of Australian mining giant Rio Tinto, and which sprawls over 6,500 acres about fifteen miles south of town. Entering a big strip mine is like visiting the supersize world of the Brobdingnags: Each mining pit (there are five at Cordero) is big enough to bury a fleet of aircraft carriers. The haul trucks make ordinary pickup trucks look like Tonka toys. As you travel down into a pit, you descend through striations of yellow sandstone and gray mudstone, moving back in time 50 million years, until you confront a black wall. "Buried sunshine," the coal industry's PR wizards like to call it.

From the bottom of the pit at Cordero Rojo, however, the view of America as the land of eternal coal is not quite so clear. After all, it's one thing to have 42 billion tons of coal buried in the ground; it's quite another to get it out. I saw this for myself one afternoon when I accompanied a twenty-eight-year-old shift foreman named Travis Todd on his rounds at Cordero. Every day, Todd tours the five pits at Cordero, looking for signs of trouble. Most often, his eyes are on the highwall — the 240-foot-high cliff of mudstone and sandstone that rises above the coal seam. Thirty years ago, when corporations began mining in the basin, the coal was buried only twenty feet or so below the surface — you could practically dig it out with a spoon. But because of the way the coal seams slope downward across the basin, all the easy stuff is gone, and the coal is buried progressively deeper with each passing day. Now the pit walls at most of the big mines are sheer cliffs. How stable are they? How much deeper can the coal companies mine without triggering collapses? Given the downward slope of the coal seam, it's not hard to see that ten or twenty years in the future, mining coal in the Powder River basin will be an increasingly complex, expensive, and dangerous operation. And if this is happening in the thickest, richest coal seam in America, what's going on in places such as West Virginia and Pennsylvania, where all the easy-to-get coal was mined out by the 1950s?

As a precautionary measure, engineers at Cordero Rojo have cut back the highwalls in some pits with a series of catch benches, or steps. It looks a little like the terraced walls of Machu Picchu. The idea is to stop or slow a collapse before it reaches the bottom of the pit, perhaps burying the men and equipment working below. These dangers are not theoretical. In 2002, at Arch Coal's nearby Black Thunder mine, a section of highwall fell, crushing a bulldozer and killing the man operating it. A month earlier, at the same mine, a rock tumbled down off the highwall and hit a pickup truck passing below, leaving the driver paralyzed.

Part of Todd's job is to spot potential danger signs and get them fixed. On a trip into the north pit, we idled along in his truck at the bottom of the coal seam while Todd scanned the highwall, looking for any cracks or signs of instability. You could see sections where big chunks of mud and sandstone had fallen away. He pointed to a thin stream of water trickling out about two hundred feet up, a sign of possible collapse. "That's something we don't like to see," Todd said, making a note of it in his logbook.

"Do you realize we have 250 million years of coal?" President George W. Bush boasted in a speech touting energy independence in 2005. He meant, of course, 250 years' worth of coal. But it's easy to get worked up when you're talking about America's vast reserves. When Tennessee senator Lamar Alexander pushes for more funding for "clean coal," he justifies it by talking about America's "500-year supply." Similarly, when Wisconsin Energy, a large electric utility, was making its case for a new coal-fired power plant, the company argued that coal was superior to natural gas because America's gas supplies are declining, while there is enough coal to last "almost 500 years." But Peter Huber, a senior fellow at the Manhattan Institute whose views have been widely promoted by the Greening Earth Society, a pro-coal trade association, gets the prize for unbounded optimism. He has called the supply of coal in North America "essentially unlimited."

In a sense, Huber is right: Strictly speaking, there will be no shortage of coal in America in the foreseeable future, and the number of years our coal reserves will last depends on how eager we are to burn it and how much better we get at mining it. In this sense, the accuracy of the official estimate from the Energy Information Administration (EIA) — the statistical arm of the U.S. Department of Energy — that the United States has 250 years' worth of recoverable coal reserves is not important. What is important is understanding what it will take in economic, environmental, and human terms to get it out of the ground. And this is where the talk about America's 250-year supply runs into trouble.

When coal boosters like President Bush and Senator Alexander tout the virtues of America's reserves, they speak as if all that coal was sitting in a big, shiny pile in the middle of Illinois or Tennessee just waiting to be shoveled into a coal plant. Unfortunately, that's not the case. A good percentage of the coal that's left is too dirty to be burned in conventional power plants, and much of it is buried in inconvenient places — under homes, schools, parks, highways, and historical landmarks. It can be mined, no doubt about it, but it won't be cheap, and it won't be pretty.

In part our unsophisticated view of coal reserves is a legacy of coal's decline. Americans stopped burning coal for heat and locomotives stopped burning it for power not because we were running out of it, but because it was displaced by cleaner, more efficient, and more convenient sources of heat and power. There was no point in studying coal reserves because the supply of coal was never an issue: it was quite obvious to everyone that there was plenty of cheap coal left in America. The problem was, would anyone want to burn it?

Now we do. And now it's worth asking some questions about that 270-billion-ton bounty of fossil fuel: Where exactly is all this coal? Is it anywhere near the power plants that burn it? Are there big differences in quality? How expensive will it be to get it out of the ground? How dangerous? What will America look like when we're finished mining it?

U.S. COAL FIELDS

These are not questions that coal boosters are eager to answer. In an era of declining resources, the great bounty of coal has always been one of the industry's best arguments for burning more. In this sense, America is more like "the Saudi Arabia of coal" than the coal industry cares to admit. Just as the last thing the Saudi oil ministers want is a true accounting of their remaining oil reserves, the last thing Big Coal wants is a close accounting of what it will take to get 270 billion tons of coal — or even 50 billion tons, for that matter — out of the ground.

"You hear the industry guys talking about it in private; they know it's not a good situation," Nick Fedorko, the head of the coal unit at the West Virginia Geological and Economic Survey, told me. "But they really don't want to talk openly about it. Mining requires a huge amount of capital investment, and if investors begin to believe that there is not as much easy-to-get coal in the ground as they have been told, then they might lose confidence. And if they lose confidence, the whole business might begin to slide downhill." Which is exactly why nobody in the industry is in any big rush to correct the myth of eternal coal in America. "People in the industry find comfort and security in these exaggerations," explains Marshall Miller, a respected geologist and coal industry consultant. "The coal industry is fighting so many battles on so many fronts — this is their one ace in the hole, and they are determined to keep it."

Coal is a creation of the earth's recent past, a legacy of the march of life from the sea onto the land. For most of the planet's 4.5-billion-year history, the earth was a hot, barren rock covered with exploding volcanoes and briny seas. Only about 400 million years ago did the first primitive life forms, multicelled algae, appear in the oceans. Fifty million years or so after that, those algae moved onto the land, evolving into mosses and, eventually, primitive plants and trees. It is the remains of this period, known to geologists as the Carboniferous Period, that created most of the fossil fuels we have today, including the great Appalachian coal bed that stretches from Pennsylvania to Alabama.

During the Carboniferous, the seas rose and fell in glacial cycles, flooding the swamps along what is now the Atlantic coast with seawater for millions of years during warm periods, then retreating again as the climate cooled and ice formed. Odd trees evolved in these swamps. The two most common types were lepidodendron and sigillaria. *Lepidodendron* trees were up to 175 feet tall and 6 feet in diameter at the base, with a few short, wispy branches at the top. *Sigillaria* trees were short and stout, 6 feet in diameter but only about 18 feet tall — more like big, leafy stumps than trees. When these trees died, they created vast mats of organic matter that were periodically covered with seawater, which cut off the oxygen supply and prevented organic decay. The creation of these swamps was further aided by an evolutionary quirk: 350 million years ago, the creatures that normally help break down the complex carbon compound in tree bark — herbivores, termites, and certain bacteria — hadn't evolved yet. With a temporary immunity from organic decay, these great mats of plant matter just kept piling up.

The creation of coal is a function of heat, pressure, and time. Plant matter is first transformed into peat — a mix of bark, leaves, roots, spores, and seeds that looks like chewing tobacco. But peat that remains near the surface will never become coal — it needs to be buried and squeezed, which creates heat and forces off the oxygen, hydrogen, and nitrogen and concentrates the carbon. Coal is divided into four ranks: lignite, sub-bituminous, bituminous, and anthracite. *Lignite* is the closest to peat, and it looks a lot like black dirt — you can still see bark and woody fragments in it. Lignite is only about 60 percent carbon. When lignite is cooked and pressurized a little longer, it turns into *sub-bituminous* coal, which contains more carbon and fewer impurities. *Bituminous* coal is even further refined. It contains about 85 percent pure carbon and is hard, flinty, and black. The highest rank is *anthracite,* which is comparatively rare and remarkably beautiful — glassy and iridescent. It is almost pure carbon and burns with a clean blue flame. If anthracite is cooked and compacted still further, it becomes graphite (not diamonds, as legend has it).

This ranking system doesn't do justice to the complexity of coal. To a coal connoisseur, coal seams are as identifiable as grape vintages. The composition of coal can change because of sudden geologic events: the eruption of a volcano can add ash and mercury; a sudden flood can inject impurities into a section of coal that makes it completely different from the coal located one hundred yards away. A chunk of coal contains virtually every element in the periodic table, and the ratio of these elements — particularly sulfur and heavy metals — to carbon is important in determining not only how the coal burns in the boiler of a power plant but also how much pollution it gives off. A ton of bituminous coal from Illinois might easily have ten times as much sulfur as a ton from West Virginia. Burning sulfur creates sulfur dioxide, which fouls the air and poisons people's lungs, making coal with a high sulfur content far less desirable for most power plants. Even relatively benign components such as sodium are important in the marketability of coal. For instance, when sodium is burned, it becomes a snotlike goo that fouls boiler tubes and drives engineers nuts.

Most of the coal formed during the Carboniferous is bituminous. The only sizable tracts of anthracite are in northern Pennsylvania, which took the brunt of the impact when the North American and African continents collided several hundred million years ago. The impact folded the mountains up like an accordion and drastically compressed the coal beds that lay within. Although clean-burning anthracite was in much demand for heating homes and powering locomotives in the nineteenth century, it's not mined much anymore, in part because the geography of the region, with coal contorted in deep zigzagging seams, makes it prohibitively expensive to extract.

In the West, coal is much younger and softer. It was created only about 55 million years ago, during the late Paleocene Epoch. At the time, North America was still recovering from the impact of a giant asteroid on Mexico's Yucatán peninsula. The asteroid had not only killed off the dinosaurs but also devastated forests and emerging flora. The first plants to recolonize the earth were huge ferns, which soon gave way to modern-looking trees, including relatives of the bald cypress and dawn redwood. At about the same time, in a dramatic geological event known as the Laramide revolution (geologists still aren't sure what triggered it), the Rocky Mountains jumped out of the flatlands. The pressure of this upward thrust caused low-lying basins to flex and sink and created a broad network of wandering streams and lush, soggy bottomland in what is now northeastern Wyoming and southern Montana. The climate was hot and wet. Redwood and cypress trees thrived in the swamps, creating vast peat bogs. Water and mud ran off the rising Rockies, periodically covering the swamps, but they persisted for 10 million years, ending only with the abrupt drying and cooling of the climate that marked the beginning of the Eocene Epoch.

Most Paleocene coal, such as that found in Wyoming, is sub-bituminous or lignite. This younger coal is much lower in heat value than the higher-ranked coal in the East. But because most western coal was created in freshwater swamps, it is also generally lower in sulfur than eastern coal, a stroke of geological fortune that has given western coal a big advantage in the energy market. In addition, whereas the best eastern coal seams are eight to ten feet thick, the coal beds in the Powder River basin are ten times thicker — and that's just around the edges of the basin. Out in the middle, lurking several thousand feet below the surface — far out of reach of today's coal miners — is an immense coalfield known as "Big George." From drilling samples, geologists estimate Big George to be two hundred feet thick — one of the single richest coal seams in the world.

America's great bounty of coal was no secret to early settlers. Unlike petroleum or natural gas, which pools in reservoirs deep underground and migrates through fissures and fractures, coal rises and falls with the folds of the earth in predictable patterns. Often it literally outcrops to the surface (where it can catch fire, burning underground for hundreds, if not thousands, of years) and can be harvested with even the most primitive tools. That's one reason coal was put to use as early as the 1300s by Hopi Indians to cook and to bake clay pottery, while oil remained a mystery until the industrial revolution was well under way.

In 1673, the French explorer Louis Joliet and the missionary Father Jacques Marquette, the first Europeans to travel down the Mississippi River, saw coal in the river bluffs near what is now Utica, Illinois. In the 1750s, a Philadelphia mapmaker who was surveying the Ohio River valley noted that along with other riches in the valley, "coal is also in abundance and may be picked up in the beds of the streams or from the sides of the exposed hills." Anthracite was discovered in eastern Pennsylvania in 1790, legend has it, when a hunter named Necho Allen was camping for the night under a rock ledge in the mountains. After building a campfire, Allen fell asleep, then awoke with alarm because "the mountain was on fire." Later, he realized he had built his campfire on a coal outcrop. Out west, coal was equally apparent. In 1874, a newspaper reporter who was traveling with the U.S. Cavalry as they chased Crazy Horse across northern Wyoming noted that "the entire section of the valley is a huge coal bed, one of the most extensive in America . . . Some day, I thought, when the Sioux are all in the happy hunting grounds, this valley will rival the Lehigh of Pennsylvania."

Because America's supply of coal was assumed to be so vast, the first attempt to quantify its reserves didn't come until 1909, when the coal industry was near the peak of its powers, and the question naturally arose how long the supply of this fuel that was so crucial to industrialized life would last. Geology was a rough art in those days, accomplished mostly on horseback and based as much on intuition as hard science. But after much study, geologists Marius Campbell and Edward Parker of the U.S. Geological Survey (USGS) estimated that the United States had a little over 3 trillion tons of coal. Of that, about two thirds, or a little less than 2 trillion tons, was "easily accessible or mineable under current conditions." This was music to the ears of early industrialists, of course, and went a long way toward establishing the dream of eternal coal in the American psyche.

The 1909 study was updated over the years, but it wasn't superseded until 1974, when Paul Averitt, also a geologist at the USGS, published a revised estimate of America's coal reserves. By the early 1970s, coal was viewed very differently than it had been in 1909: it was clear that coal's glory days were in the past. In addition, America was facing its first energy crisis, and politicians and business leaders were faced with difficult strategic decisions about the country's energy future. The subtext of Averitt's study is modest and practical: Does America have enough coal to last until it can be supplanted by nuclear power, solar, and other renewables?

Although Averitt used data from the 1909 study in his paper, he also had access to a vast amount of new information about coal that had been accumulated by mining and oil-drilling companies. In addition, he considered factors that had been ignored in the 1909 study, such as the thickness of the coal seams, the amount of earth that needed to be moved to get to the coal, and the general quality of the coal. His conclusion: the amount of coal that was suitable for mining — what he identified as "the reserve base" — was only about 483 billion tons. When you considered losses due to mining operations, the rank and quality of the coal, and other restrictions, Averitt estimated that only about 50 percent of the reserve base, or about 243 billion tons, was recoverable. That was a long way from the 2 trillion tons in the 1909 study, but it was hardly alarming: at the rate America was burning coal in 1973, there was still enough easy coal left in America to last four hundred years.

Almost before the ink on Averitt's study was dry, it became clear to many geologists that Averitt hadn't gone far enough. Among other things, he had failed to make allowances for the sulfur content of various coals, which, since the Clean Air Act was passed in 1970 limiting sulfur dioxide emissions from power plants, had a huge impact on the coal markets. In high-sulfur coal regions such as Illinois, Ohio, and central Appalachia, demand plummeted; in low-sulfur regions such as Wyoming, it boomed. Clearly, the market was far more subtle than the science.

In 1986, the USGS began a pilot project with the Kentucky Geological Survey to attempt a more precise accounting. Instead of surveying the entire state, the USGS focused on one quadrangle (about sixty square miles) in the Matewan coalfields of southeastern Kentucky, one of the most historically important regions for the coal industry in the state. Geologists looked closely at the real-life restrictions on mining: state and national parks, roads, towns, proximity to railroads, coal quality, losses during mining operations, and geological limits such as rock intrusions and narrow coal seams. The result: of the 986 million tons of coal in the quadrangle, only about 30 percent of it was potentially mineable.

Over the next couple of years, the USGS studied about twenty other quadrangles in West Virginia, Kentucky, and Virginia. They all showed similar results: when you factored in obvious restrictions, far less than 50 percent of the coal that Averitt had estimated as "recoverable" was really available for mining. But these studies, comprehensive as they were, still left out one key factor in determining how much coal was really available: economics. Nobody mines coal for fun, after all. The important question is not how much coal America has, but how much of it can be mined at a profit given market prices.

In 1989, the USGS brought in the U.S. Bureau of Mines, an expert at the engineering and economics of coal mining, to add yet another layer of reality to the coal assessments. The bureau started again with the Matewan quadrangle in Kentucky. Headed by a young geologist and engineer named Tim Rohrbacher, the bureau's team used census data, mine maps, and other information to determine that at $25 a ton (about the price of coal when the study was completed), only 7 percent of the coal in the quadrangle was economically recoverable; at $30 a ton, 22 percent was economically recoverable. At the conclusion of this study in 1993, Rohrbacher warned that the Matewan area could be mined out in as little as seventeen years. "If similar results are found in subsequent investigations," Rohrbacher wrote, "a strong argument can be made that traditional coal producing regions may soon be experiencing resource depletion problems far greater and much sooner than previously thought." In later studies of coalfields from Illinois to Colorado, Rohrbacher came up with remarkably similar results: only about 5 to 20 percent of original coal resources were available for mining at current prices. Even in the mighty Powder River basin, in the quadrangle that Rohrbacher examined, only 11 percent of the coal was economically recoverable.

Rohrbacher, who is now employed by the USGS, is at work on a national reassessment that will be completed in five to ten years. But the work he's published so far underscores an obvious truth: the often-quoted estimate that we have 250 years' worth of cheap coal left in America is a gross exaggeration. Rohrbacher points out that the EIA's optimistic estimate of recoverable coal reserves is still based on Averitt's 1974 study, which is not only more than thirty years old but also is itself based on reports from state geologists that are fifty to sixty years old. Averitt's numbers have been massaged a little by statisticians, Rohrbacher says, "but statisticians are like car salesmen — first they ask you what number you want, then they figure out a way to come up with it."

Even the EIA admits the numbers are imprecise. As it turns out, there is exactly one person, an overworked sedimentary geologist named Rich Bonskowski, who is in charge of coal numbers at the EIA. He admits that it bothers him when he hears coal industry executives — not to mention the president of the United States — talking publicly about 250 years' worth of coal reserves in America. "I'd like to be able to bring sober assessment to those numbers," Bonskowski explains, "but we just don't have the resources." Not surprisingly, coal lobbying groups are even looser with their facts than the EIA. Among the most notorious is the West Virginia Coal Association. The EIA claims there are 33 billion tons of coal left in the state; the coal association boosts it up to 53 billion tons. Both numbers, says Nick Fedorko of the West Virginia Geological Survey, are "fraught with problems." A more realistic accounting of recoverable reserves, Fedorko says, is 9 billion to 19 billion tons. "And even that," he admits, "may be optimistic."

In the 1970s, the depletion of the earth's resources was widely predicted, and a lot of smart people ended up looking pretty silly when they failed to anticipate how good engineers would be at figuring out ever more efficient and economical ways of getting coal, oil, and gas (among other things) out of the ground.

In the early part of the twentieth century, coal mining progressed largely by replacing men with machinery: the picks and shovels were banished, replaced by mining machines that did twice the work in one third the time. During the past thirty years, increased productivity has largely been driven by the speed and size of the machinery: draglines — huge, cranelike earthmovers with long booms and buckets large enough to pick up a small house — and haul trucks have gotten bigger, conveyor belts have gotten faster, and coal trains have gotten longer and heavier. Bigger, better equipment has allowed coal companies to expand the size of their operations. In Wyoming, as well as in parts of Appalachia, mines now sprawl over thousands of acres. This new equipment also gave birth to mountaintop removal mining. Instead of tunneling after coal, which has always been slow, dirty, dangerous work, it became more economical to strip away the mountain; haul out the coal with giant trucks; and then pile up the old dirt into something resembling what was there before, toss some grass seed on it, and call it a job well done.

Big Coal's most valuable tool is not bigger, better haul trucks, however. It's ANFO — ammonium nitrate/fuel oil. ANFO is a high-tech explosive that is used in big strip mines all over the world, as well as by terrorists such as Oklahoma City bomber Timothy McVeigh. Instead of digging out the coal, it's much more efficient to blast the earth around it away. Coal mines are responsible for about 70 percent of the 2.5 million tons of industrial explosives that are detonated in America each year. At times, there are so many blasts going off in the Powder River basin that the region could easily be mistaken for an air force bombing range. Blasting engineers like to tell a story about how Russian scientists picked up suspicious tremors on their seismographs a few years ago and accused the United States of secretly testing nuclear weapons in northeastern Wyoming. No, our diplomatic staff explained, we're just mining coal.

At Cordero Rojo, I asked the blasting supervisor, Gary Jerke, if I could watch an ANFO blast, and he happily agreed. Jerke is stocky and broad-shouldered, with a salt-and-pepper beard and a scuffed white hardhat that is plastered with old stickers from blasting product manufacturers. One morning, we pulled up on a broad, flat area above one of the mining pits, where orange detonation cord zigzagged over dusty gray mudstone. Beneath the cord, in an area roughly the size of a football field, about 180 holes had been drilled in the ground, each fifty feet deep and ten inches wide. Each hole had been filled with ANFO the previous afternoon and was ready to go. Timothy McVeigh used only 3,800 pounds of ANFO to blow up the Murrah Federal Building in Oklahoma City. Buried here was 55,000 pounds. By Wyoming standards, it was a small load.

Jerke immediately began arming the remote-control detonator. Helping him was Paul Hazlet, also a blaster at the mine. "We're trying out a new kind of detonator today," Jerke explained. He connected the end of the blasting cord to the detonator box, then opened it up and showed me a small steel pin inside. "When I push the button, that pin makes a spark, which ignites the blasting cord."

The whole explosion is elaborately choreographed, with a 130-millisecond time delay spliced into the cord between every two holes. "The goal here," Jerke told me, "is not to blow the place to hell. We just want to fluff up the dirt a little and make the coal easier to dig out."

Jerke and Hazlet double-checked the connections to the detonator, then closed a plastic lid and set it on the ground. "Okay, we're armed and ready," Jerke said.

Jerke handed a small orange box with an antenna on it to Hazlet. It looked like something you might use to fly a model airplane. Then Jerke drove off to observe the blast from a safe distance. I jumped into another truck with Hazlet, a friendly, rugged-looking guy with red hair and ragged coveralls. As he drove, he told me he'd been a blaster in Wyoming for ten years. "Still got all my fingers and toes," he joked. The detonator sat on the seat between us like a box of candy.

We parked on a level spot overlooking the blast site and stepped out of the truck. Hazlet handed me the orange box. "All you have to do is flip the switch," he explained. "But not until I tell you to."

Hazlet radioed Jerke and told him we were in position. There was much radio chatter as Jerke checked with everyone to make sure the roads and the blast area were clear. Then a warning siren sounded.

Hazlet reached over and turned a key on the box. A light blinked. "Okay, it's all set," he told me. "When you hear 'Fire in the hole,' throw the switch."

Jerke on the radio: "Thirty seconds to blast."

I stared at the box, my finger poised on the toggle.

"Don't look down when you throw the switch, or you'll miss it," Hazlet warned.

I looked up.

"Fifteen seconds to blast."

I wondered, briefly, if I would feel bad about doing this.

Jerke: "Fire in the hole."

I flipped the switch.

The explosion arrived in three parts. First, the earth lifted like a giant blanket being shaken out. Second, the ground beneath my feet trembled. And finally, I heard a surprisingly muffled boom.

After the dust settled, Hazlet and I drove back down into the pit. I was buzzing with adrenaline — cracking open the earth certainly perks you up.

Soon Jerke and others arrived. We inspected the devastation — large continents of mudstone strewn about, lifted and fractured, broken up. A job well done, Jerke proclaimed. Within an hour, a big shovel would move in and begin uncovering the coal. Within twenty-four hours, that coal would be loaded into a railroad hopper and rolling over the prairie toward a big coal burner somewhere in America, where it would, as Jerke put it, "feed the beast."

That morning at Cordero Rojo, I understood why some people are so optimistic about coal's future. The amount of coal we can extract from the earth is limited only by the size and frequency of the ANFO blasts we're willing to detonate. As long as we're willing to blow the hell out of the place, there is no coal seam, no matter how remote or how deeply buried it may be, that can't be mined.

If all of America looked like the Powder River basin, it would be hard to mourn the loss of a few square miles of prairie in exchange for cheap, abundant energy. Unfortunately, the Powder River basin is a unique place. In the rest of the country, ANFO blasts are more like artillery shells in an industrial war. And the people who are getting the worst of it aren't the lobbyists or legislators in Washington who push for more coal burning. They're the people who were unlucky enough to have built homes and raised families above the remaining coal seams.

No matter how you cut it, coal mining is vastly more invasive than oil or gas drilling. Done right, inserting a drill bit deep into the earth to tap a pool of oil or gas is minimally destructive. But mining coal is always brutal. It's hard on people and it's hard on the land. In some cases, this brutality can be a virtue: in a big strip mine, you can extract a lot of energy per acre of land disturbed. And some coal companies, including Kennecott Energy, which owns Cordero Rojo, have become much more sophisticated about how they reclaim land after they're finished mining. But coal mining, by its very nature, is a destructive, dangerous operation that does not fit well with civilized life. Just ask the parents of three-year-old Jeremy Davidson.

On the night of August 20, 2004, his parents tucked him into his crib in their small house in Inman, Virginia. A short time later, a bulldozer working at a strip mine above them pushed a six-hundred-pound boulder over an embankment. It kept rolling, bounced down the hill, and crushed Jeremy as he slept.

Fifty years ago, children like Jeremy didn't sleep immediately below big strip mines. Most coal mines were underground, and they were located in remote areas, keeping the messy reality of mining operations far removed from most people's daily lives. But dwindling reserves, combined with expanding population, are changing that. In Pennsylvania, miners now routinely tunnel under roads and houses, causing the land above to subside, breaking windows, cracking foundations, fraying nerves, and sometimes rendering houses uninhabitable. In eastern Ohio, Bob Murray, a well-known and politically connected coal operator, sparked a fierce battle with local residents over his plans to mine beneath the most significant old-growth forest in the state. In Pike County, Kentucky, dozens of families were recently evacuated from their homes because of the danger of flying rocks caused by blasting at a strip mine nearby. In West Virginia, overloaded coal trucks terrorize small towns and have been involved in numerous fatal accidents.

If America's dependence on coal grows, so will these kinds of conflicts and confrontations. The coal industry tends to see people who live near mineable coal seams as people who are standing in the way of progress, as if their right to profit from digging out these ancient black rocks supersedes every other right, including the right of future generations to imagine what the Appalachians looked like before they got leveled. The issue is not really whether we have enough coal to provide enough electricity to keep our air conditioners cranked up. We surely do. The issue is, how big a part of America are we willing to sacrifice for this privilege?

In West Virginia, residents can see the topography of the state changing before their eyes. Not long ago, I spent an afternoon with Larry Gibson, a local resident whose home is on a mountaintop that is surrounded by a big mountaintop removal mine. Draglines and shovels have already carved away much of the land that surrounds his property, which has been in his family for three hundred years, leaving him living in a small oasis of trees in the middle of an industrial zone. Streams where Gibson's great-grandfather fished and swam are now filled in; hills he hiked with his father are flattened. Gibson, understandably, has taken it personally and keeps a .38 under the seat of his truck to defend himself against angry coal miners who don't like his very vocal objections to the way they conduct business. "People say, 'Larry, what are you so angry about? America has to get electricity from somewhere, doesn't it?'" he explained. "Then I take them out here and show them this, and they understand."

Gibson and I stood together on a bluff near the family graveyard, where we could look down into the vast terraced mining pit below. A distant boom echoed through the hills; the ground rumbled, and dust plumes drifted over the trees. I recognized the sound. The mountains were coming down, one load of ANFO at a time.

Chapter 2

Coal Colonies

EARLY ON A COLD, rainy December morning, sleepy but excited children and their sleepy but less excited parents began to line up outside an old brick furniture store in Madison, West Virginia. Madison is a small town of about 2,600 on the Little Coal River about an hour southwest of Charleston. The town is best known as the gateway to the southern coalfields, once called West Virginia's billion-dollar coalfields because they were so vast and so rich. The legacy of Big Coal is visible everywhere here, from the statue of the coal miner in front of the Boone County courthouse to the thin layer of black dust that coats the railroad tracks on Main Street. Southern West Virginia is still the most productive coal region in the state, but its glory days are long past. Like most coal towns, Madison feels like a harsh, unforgiving place, where just getting by is an accomplishment.

But not that morning. That day, the biggest coal company in West Virginia, Massey Energy, was holding its 2004 Christmas Extravaganza. Madison is practically a company town for Massey — several of its biggest strip mines are in the hills nearby, and the $3.1 billion company employs some eight hundred workers in the area, making it by far the largest employer. Massey employees and their spouses had spent most of the last week getting ready for the Christmas Extravaganza, brightening up the old furniture store with a Christmas tree, tinsel, and holiday cheer. Santa Claus would be there, along with his elves, passing out free toys to children and free frozen turkeys to their parents. Massey had deliberately targeted the poorest of the poor in the region, sending home flyers with children who were on the free lunch program at the local schools. "Dear Family," the flyer began, "Massey Energy Company is having the Second Annual Christmas Extravaganza and your child is invited!" About five thousand of these flyers had been distributed. "This is our way of giving back to the community," a Massey spokesperson told a local reporter. "For some of these kids, this is the only Christmas they'll have."

By the time the doors opened at 11:00 A.M., the line of wet, cold families was nearly a block long. People began filing in, passing a pair of security guards, and were greeted by cheerful Massey employees and their spouses in red shirts and Santa Claus hats. The eager, bright-eyed kids were guided through aisles of toys — dolls, portable CD players, toy soldiers, miniature NASCAR racecars, stuffed animals. They were urged to take two toys each, and only two, and keep moving. And they did, gratefully and happily. It was an assembly line of Christmas joy.

At about noon, a big green and white customized bus with tinted windows pulled up beside the old brick building. It was the kind of vehicle a country music star might travel in. But when the door opened, it was not Kenny Chesney who emerged — it was Don Blankenship, the CEO of Massey Energy. He was dressed casually, in a soft leather bomber jacket and the same red Massey Energy Christmas Extravaganza shirt that all his employees had on. As he made his way toward the party, heads turned — even the kids in wheelchairs knew who he was and, more important, that he was one of the richest men in West Virginia. His compensation in 2004 alone had been more than $6 million, making him the highest-paid executive not only in the state but in the entire coal industry. For Blankenship, the holidays had come early: just weeks before the Christmas Extravaganza, he had cashed in Massey stock worth $17 million. This was all the more remarkable considering the fact that Massey was hardly a cash cow. During the previous three years, the company had lost at least $60 million.

Blankenship's political power was as impressive as his wealth.

Massey's board of directors was made up of some of the most influential players in West Virginia business and politics, including James H. "Buck" Harless, a coal and timber baron who had raised hundreds of thousands of dollars for the Bush campaign in 2000 and 2004. (After serving four years on Massey's board, Harless resigned in early 2005.) During the 2004 election, Blankenship had spent $3.5 million of his own money — a huge amount in a small state like West Virginia — on political attack ads to unseat a state supreme court justice who had ruled against Massey in a civil case, potentially costing the company millions in damages. A few months later, Blankenship spent another half million or so on ads to defeat Democratic governor Joe Manchin's controversial proposal to refinance West Virginia's crippling public pension debts. When the governor dared to suggest that Blankenship's high-profile involvement in state politics might lead to closer scrutiny of his coal business, Blankenship promptly sued the governor, accusing him of violating his right to free speech by threatening to selectively enforce state laws against him. One admiring conservative columnist tagged Blankenship, with unintended irony, "the ultimate twenty-first-century coal baron."

Despite his wealth and power, Blankenship is not an imposing figure. He is fifty-six years old, with dark thinning hair, a mustache, and a modest potbelly. If you didn't know better, you might mistake him for a moderately successful Chevy salesman. But his understated manner — "I'm a poor guy with a lot of money," he likes to say — and his folksy drawl are misleading. According to one former Massey executive, Blankenship kept a TV with a bullet hole in it in his office for years, a souvenir of the labor battles of the early 1990s and a not-so-subtle reminder to guests that he was not easily intimidated.

Blankenship, who often identifies himself as a born-again Christian, comes from an old Appalachian family that arrived in Kentucky in the mid-1800s. The Blankenships settled near what is now the tiny town of Stopover, which is just a mile or so across the Tug Fork River from Mingo County, West Virginia. Blankenship's mother, Nancy McCoy, is a descendant of the clan renowned for their bloody battles with the Hatfields back in the 1880s. Blankenship never knew his father. In the early 1950s, while his mother was raising three older children in Stopover and her husband was away in Korea, she got lonely, met someone else, and became pregnant with Don. Her marriage broke up, and she and her infant son wound up on a bus, eventually settling in Mingo County (where Blankenship still lives). She was later joined there by her three older kids and used money from her divorce settlement to open a roadside grocery store and gas station. In Blankenship, hard times bred ambition. "I knew I had to do something because I had nothing to fall back on," Blankenship has said, speaking of his childhood. "I knew I wasn't getting off to a real good start." After graduating second in his high school class in 1968, Blankenship attended Marshall University in West Virginia, where he studied accounting. He spent ten years working in the food industry — first for a big cookie company, then for a bakery that delivers packaged bread throughout the Southeast — before he grew homesick for Appalachia. In 1982, Massey called and offered him a job in coal sales. Ten years later, he was running the company.

Not long after Blankenship arrived at the Christmas Extravaganza, someone handed him a red Santa hat, which he pulled down over his balding head. He looked ridiculous, of course, but who cared? It was Christmas. He stepped into the old furniture store and spent the next few hours hobnobbing with employees, greeting children, posing for photographs, and generally basking in good cheer. Amid so much poverty, there was an air of magic around Blankenship. Kids came up and thanked him for their toys while their parents stood back and watched, as if hoping that his good luck would rub off on their children. To many people in Madison, Blankenship was the local boy made good — a scrappy, slouching hillbilly who had climbed to the top of a tough industry and was now a bona fide millionaire. That's the American dream, West Virginia style.

Outside in the drizzling rain, Maria Gunnoe and her ten-year-old daughter waited in line to enter the party. Gunnoe was a thin, wiry thirty-six-year-old woman with long, dark, wavy hair and a habit of speaking her mind. She lived up-hollow in a little place called Bob White, a few miles south of Madison. Like most of the people at the event, she was from a coal mining family — her grandfather, father, and brothers all worked in the mines. Unlike nearly everyone else in line, Gunnoe was not a fan of Don Blankenship. In fact, she believed he was pretty much single-handedly responsible for destroying her beloved West Virginia.

Gunnoe came to her views the hard way. In the spring of 2003, Big Branch Creek, which ran only a few hundred feet from her house and was usually small enough to jump over, became a wall of black water roaring down out of the hollow. In the fifty years her family had lived in Bob White, nothing quite like that had ever happened before. Rocks the size of Volkswagens tumbled down the river. The force of the water yanked Rowdy, her rottweiler, right out of his collar and carried him off. Gunnoe dashed through waist-deep water to fetch her daughter at a neighbor's house, then carried her back through the rising current. She believed they would both drown. Somehow they made it through, and Gunnoe and her family spent the night huddled in her little house above the Big Branch, wondering if the water would wash them away.

Until that moment, Gunnoe had never quite grasped the consequences of the big new strip mines that had opened in the hills above her in 2001. She had heard the blasting and swerved out of the way when the coal trucks came barreling around the corner on one of the local roads. It was scary, but she'd dealt with it. Then the flooding began. In three years, Gunnoe was flooded six times. It was no mystery what was happening: as the mountains above her were disassembled, the rock and debris was dumped into the headwaters of creeks and streams, creating what the coal industry innocuously calls "valley fills." When it rained, the naked mountains guttered the water into the hollows. The filled-in headwaters of the creeks only accelerated the momentum of the runoff during storms, often turning a small, docile-looking creek like the Big Branch into a raging torrent. This was not a problem particular to Bob White. More than seven hundred miles of streams had been filled in throughout Appalachia, changing the natural drainage patterns and making catastrophic flooding a springtime ritual in the southern coalfields.

Even more dangerous, Gunnoe realized, were the big slurry impoundment ponds that are often built at mining sites — huge man-made lakes designed to store the runoff from coal washing, which are often filled with sludge containing high concentrations of heavy metals such as lead, arsenic, and selenium. In heavy rains, the earthen dams that hold these impoundments back sometimes fail, sending tidal waves of black, polluted water down over the people living in the hollows below.

The floods woke Gunnoe up to what was happening around her. It wasn't just the blasting of the mountains and the floods in the hollows, she realized. It was the destruction of a whole way of life. It was the fish that were gone from the streams, and the startling number of people she knew who had been diagnosed with cancer, and the kids with asthma, and the slurry impoundment ponds that leached chemicals into the drinking water. And most of all, it was the hopelessness and fear she saw all around her. Whenever she pulled up at a gas station in Boone County and saw a man with a particular look of sadness and desperation in his eyes, she wondered what coal company he worked for and whether more than one hundred years of taking orders from mine superintendents and coal barons had crushed something essential in West Virginia's soul.

Gunnoe knew that Don Blankenship was not responsible for all of this, that there were other coal companies in the hills above her, and they were in some ways just as bad. But to Gunnoe, no one embodied the lawlessness and power of Big Coal as much as Blankenship. Blankenship had made his name as a union buster and had done more than any other man to lower the wages and cut the benefits of miners in the state. Even by coal industry standards, his disregard for state and federal mining regulations was remarkable. Between 1995 and 2000, Massey Energy had been cited in 531 separate enforcement actions by state and federal agencies. The Martin County slurry spill in 2000, which had released 300 million gallons of sludge into Appalachian streams, had occurred at a Massey-operated impoundment in Inez, Kentucky. The spill covered seventy-five miles of the Big Sandy River with black sludge, killing 1.6 million fish, washing away roads and bridges, and contaminating the water systems of more than 27,000 people. Incredibly, no one had been killed. The Environmental Protection Agency (EPA) called it the largest environmental catastrophe in the history of the southeastern United States.

Massey's safety practices were equally troubling. In July 2001, the United Mine Workers published a report that alleged Massey had "the nation's worst fatality record" among coal companies. Five workers had died at Massey mines in a ten-month period beginning in April 2000, the report charged. A few months later, an investigation commissioned by West Virginia's then governor Bob Wise concluded that Massey's safety record was confused by the company's extensive use of subcontractors at its mines. "Massey Energy has for several years publicly proclaimed that their accident record is among the lowest in the nation," the governor's report said. "However, if contractor accident data were included, it would be among the highest." Although Massey claims to have improved its safety practices lately, the deaths of two miners after a fire in Massey's Alma mine near Melville, West Virginia, in January 2006 demonstrated that the company still had a long way to go.

To Gunnoe, the fact that Blankenship was a local boy, and not a faceless corporate bean counter, made his crimes even harder to bear. It was one thing for an outsider to come into a place like West Virginia and cut up the mountains and exploit the people. It was quite another for a person who had been born and raised here to do it, and to be so shameless and disrespectful to his own kin that he thought they could be bought off with a few Christmas trinkets that he passed out once a year.

Standing in line that morning, Gunnoe felt the pull of conflicting emotions. Her fourteen-year-old son had refused to come, but her daughter, who was too young to understand the ethical complexities of the situation, had begged Gunnoe to bring her down to the Christmas Extravaganza for a free gift. And why not? Gunnoe was no better-off than most people in this line. Besides, she thought of the whole experience as a kind of guerrilla operation. She had printed up a couple of hundred flyers listing Massey's many sins and, with the help of a friend, left them on the seats of the shuttle buses that brought people to the event.

But the closer she got to the door, the more she began to feel that coming had been a mistake. Seeing Don Blankenship surrounded by children, cameras whirling — she knew the media, especially the local Fox station, would be playing up these images all through the holidays — "I started feeling physically ill," she said later. If he really wanted to help the poor, he didn't need cameras. The kids were just props to him, she thought. She looked at his fine leather jacket, the posh bus outside, and the well-dressed minions hovering around him. This whole event, as far as Gunnoe was concerned, was nothing but a way to buff up Massey's reputation in front of the locals. *Look, we gave your daughter a little plastic doll. We really care.*

And the saddest thing of all, she thought, was that people were falling for it.

In the beginning, the coal colonies all had big dreams, especially West Virginia. Even before the state joined the Union in 1863, many West Virginians believed that its abundance of natural resources would make them all rich. Sure, the state was rugged and remote, its hollows settled by mountain men and muleskinners, but vast tracts of coal and timber — the very raw materials of the booming industrial economy — made the Mountain State's triumph seem inevitable. Joseph Diss Debar, a land agent and influential legislator, summed up the feelings of many West Virginians in the 1860s: "That such a country, so full of the varied treasures of the forest and the mine . . . should lack inhabitants, or the hum of industry, or the show of wealth is an absurdity in the present and an impossibility in the future."

Nearly 150 years and some 13 billion tons of coal later, it's strikingly obvious that the great wealth of natural resources in West Virginia has been anything but a blessing. Rather than bringing riches, it has brought poverty, sickness, environmental devastation, and despair. By virtually every indicator of a state's economic and social well-being — educational achievement, employment rate, income level — West Virginia remains at or near the bottom of the list. Nowhere is the decline clearer than in the southern part of the state, where the promise of riches was once brightest. McDowell County lost nearly half its population in just twenty-three years. The region has some of the highest obesity, cancer, and loss-of-teeth rates in the country. Michael Hicks, an economist who has written widely about West Virginia's troubles, says that the literacy rate in McDowell County is about the same as that in many Third World countries. "The economic conditions in southern West Virginia today are a human tragedy of epic proportions," he says.

What went wrong? For the past forty years — since April 24, 1964, when President Lyndon Johnson stood on the steps of a mountain shack in Inez, Kentucky, and declared a war on poverty — the decline of Appalachia has been the subject of debate and study. A key question is the role the coal industry has played in the economic development of the region: is it a blessing or a curse?

The question applies not just to Appalachia but to every region of the country where coal mining has prospered — Ohio, Alabama, Illinois, and even, to some degree, Wyoming. How is it that a rock that has been the basis for so much prosperity and progress in America has left behind such a trail of human and environmental wreckage in the regions where it is mined?

Of course, coal has been very good to some. The best evidence of this is the Elms, one of the grandest and most frequently visited Gilded Age mansions in Newport, Rhode Island. It was the home of Edward Berwind, an early-twentieth-century coal company owner and close friend of financier J. P. Morgan. Although Berwind is not remembered for much besides his gaudy mansion, at the turn of the century this hard-working son of a German immigrant and cabinet-maker was voted one of the thirty most powerful people in America. At the time, the Berwind-White Coal Mining Company was one of the largest coal companies in the United States, selling bituminous coal from Pennsylvania and West Virginia to the railroads, the U.S. Navy, and many steamship lines.

Completed in 1901, the Elms is a copy of an eighteenth-century French chateau built almost entirely out of Indiana limestone. Berwind and his wife, the daughter of the U.S. ambassador to Italy, filled the mansion with a jumble of European art and furniture, including Venetian murals in the dining room, dainty French neoclassical furniture in the drawing room, and half-naked nymphs and goddesses everywhere you look. The effect is suggestive of nothing so much as Berwind's faith that money could buy anything, even good taste and civilization. But the most revealing detail at the Elms may be a large bronze sculpture out on the rear terrace that depicts a battle between an alligator and a lion. The alligator is on the bottom, belly up, thrashing mightily, but it is clearly no match for the strong, agile lion, whose left paw is pressed firmly against the reptile's throat. If there's a better symbol of the power and ruthlessness of laissez-faire capitalism, I've not seen it.

Berwind's coal empire covered much of central Appalachia, but the heart of it was in Windber, Pennsylvania (which was named after Berwind, but to avoid confusion with another town by the same name, the syllables were transposed). The town is about an hour north of the West Virginia border, tucked away in the folds of the old bituminous coalfields of Somerset County. At its peak in the early 1900s, some ten thousand people lived in Windber, mostly immigrants from southern and central Europe who had been enticed by the coal company to come and work in the mines. According to a 1923 investigation, the conditions there were "worse than the conditions of the slaves prior to the Civil War." In 1906, three miners were shot, probably by Berwind's private security guards (the investigation of the murders, such as it was, was inconclusive) during a protest outside the town jail.

Berwind died in 1936, but the company continued mining coal in Windber until the 1970s. Just outside the town, you can still see the remains of Berwind's last mine in the area, Scalp Level. When I visited on a snowy day in mid-December, it felt as if I had stumbled into a lost world. The mine portal has been cemented closed, weeds growing up around it, but much of the outer infrastructure is intact. The old bathhouse is still there, the lockers dusted with soot and graffiti, the ceiling covered with hoists on which the miners hung their clothes. In the manager's office, amid shards of broken glass and beer cans, I found old mine maps and memos about fatal accidents issued by the U.S. Bureau of Mines. July 1972: "A roof bolter was so severely injured when struck on the head by a whirling roof bolt power wrench while installing roof bolts that he died on route to the hospital." November 1974: "A roof-fall accident occurred in which a jack setter was fatally injured. The victim was nineteen years of age and had six months mining experience." Whatever small measure of prosperity this mine had brought to the region, it was gone now. Somerset County is among the poorest counties in Pennsylvania. Windber is desolate, with boarded-up storefronts and rows of sagging white houses hugging each other for support.

By contrast, the Berwind Corporation is doing just fine. It has morphed into a family-owned investment management company that is, according to its promotional material, "dedicated to long-term capital appreciation." The Berwind Corporation's investments in real estate alone are valued at $3 billion. If you ask what the people of Windber got in exchange for this plundering by Berwind, the best answer is a trip to the cemetery. In the region around Windber, you will search in vain for a Berwind library, a Berwind college, a Berwind hospital, or even a Berwind park bench. Unlike steel kings such as Andrew Carnegie, who eventually spent their millions endowing libraries, museums, and universities, the coal barons didn't bother. As George Leader, a former governor of Pennsylvania, put it during a 1995 interview:

> There is something about the extractive industries that, somehow, exploitation seems to be the only word that applies. They don't seem to care about the hospitals or the churches or the community buildings or even the infrastructure unless it directly affects them. They just never did anything to help the community. They just got in and they took their money and they did as little as they could to protect the workers from dust, from cave-ins, from anything. They just did the minimum. That's what it was all about. Get in . . . get their money[,] and get out.

In theory, coal mining should be a boon for places like Madison, West Virginia. Buried underground, coal is worth nothing. Mining brings in jobs, creates infrastructure (roads, bridges, railroads), and provides spin-off benefits in everything from more jobs for maintenance workers to an increase in coffee sales at the local diner. In addition, coal companies pay a severance tax (in West Virginia, it's currently 5 percent) on every ton of coal they take out of the ground, which is a boost to the state's revenues. In 2005, the West Virginia coal industry contributed more than $238 million to the state's coffers. What's not to like?

In the early 1990s, however, economists began to explore the link between the extraction of natural resources and economic prosperity — or the lack of it — more deeply. This reevaluation was prompted by the economic troubles of many Third World countries that had recently built economies around natural resource extraction. Nigeria is a textbook example: Between 1965 and 2000, the country received $350 billion in oil revenues. By conventional economic logic, the people of Nigeria should be better-off since the discovery of oil. Instead, they are poorer than they were before oil was discovered. The country has been torn apart by ethnic and religious conflict, corruption is rife, and the educational level is low. And Nigeria is hardly alone. Chad, Sudan, the Democratic Republic of the Congo, Venezuela, and Colombia are often cited as examples of what is now known as "the resource curse."

In a landmark 1995 study, Harvard economists Jeffrey Sachs (now director of the Earth Institute at Columbia University) and Andrew Warner discovered a clear negative relationship between natural resource–based exports, including agriculture, minerals, and fuels, and gross domestic product (GDP) growth. Of the ninety-five developing countries that Sachs and Warner investigated, only two resource-rich countries achieved even a 2 percent annual GDP growth rate between 1970 and 1989. Even the oil minister of Saudi Arabia, Sheik Ahmed Yamani, viewed the presence of oil in his country with a fair degree of ambivalence: "All in all, I wish we had discovered water."

On one level, the problem in countries such as Nigeria is obvious: control over natural resources allows a few people to obtain tremendous wealth, giving them huge sway over the economic fortunes of the state and offering enormous opportunity for self-indulgence and corruption. At best, economies that are dependent on natural resources are unstable. When coal or gas prices are up, they're awash in cash; when prices fall, they struggle to keep the lights on in hospitals and gunfire from breaking out in the streets. This kind of economic yo-yoing leads to budgetary and financial fiascoes, in addition to leaving the government open to economic blackmail by the extraction industries: *If you don't let me mine that mountain, I'll pull out and leave you all in poverty.*

The deeper problem for many of these countries is that the development of natural resources tends to crowd out the growth of other, more sustainable industries such as manufacturing. Economists have come up with a number of explanations for why this is so, including the so-called Dutch disease, named after a phenomenon observed in the Netherlands during the North Sea oil boom of the

1970s. During that time, the oil and gas industry drove labor costs so high that other industries went out of business. In addition, extraction economies don't traditionally put a very high value on education. Most mining and drilling jobs require more brawn than brains, and the few highly skilled jobs are usually filled by imported workers. Because of this, investment in education is usually minimal. After all, the last thing any coal company wants is an educated, independent-minded worker who might decide to walk out of the mine and start his or her own business.

The resource curse is not inevitable. The United States and England are obvious examples of how wealth generated from natural resources can be used to build strong, sophisticated modern economies. Norway, thanks to its progressive government and a view of natural resources — oil, fish, and forestry products — as *national* resources, the benefits of which are to be shared by all, has been able to achieve arguably the highest quality of life in the world. Even some African nations have made the leap. At the time of its independence in 1966, Botswana was one of the poorest countries in the world. Now, thanks largely to the success of its diamond mines, it has one of the highest per capita income growth rates. Economists don't offer many easy explanations of why the curse clings to some countries and not others, but most would agree that two factors are crucial for success: investment in education and a transparent political process. Without those ingredients, the wealth created by natural resources does nothing for the general population. The vast riches of the Congo — first diamonds, now coltan, a rare mineral used in cell phones and laptops — built palaces in Belgium but has left the Congolese dying of famine and civil war.

In important ways, West Virginia is more like the Congo than it is like the rest of the United States. The state's economic DNA was shaped in part by its mountainous terrain, which made shipping goods in and out by railroad difficult and expensive and therefore limited manufacturing. Like many Third World countries, West Virginia's early development was greatly influenced by outsiders bent on one thing: exploiting the vast natural wealth of the state. Men like Henry Davis, a coal and railroad baron, and his son-in-law, Stephen Elkins, who was elected to the U.S. Senate from West

Virginia in 1895, helped to construct a political economy that, as historian John Alexander Williams puts it, "allowed men like Davis to identify their private interests with the public welfare and to pursue it successfully by political means." As early as 1900, it was clear that West Virginia was falling behind other states. "In [blunt] terms," Williams writes, "it developed a colonial economy and remained in the industrial age the backwater it had been in pre-industrial times."

The full tragedy of West Virginia's extraction-based economy was not apparent until after World War II. By then, the demand for coal was beginning to wane, and new machinery had improved productivity in the mines, allowing coal companies to reduce their work forces. At the same time, the rise of strip mines, with huge shovels and draglines that dug coal out of the ground and devastated mountains, created a sudden visual analogue to the economic exploitation that had been going on in the region for nearly one hundred years. Writer Henry Caudill captured the human consequences of it in the early 1960s, when he wrote about "a gray pall over the whole land . . . [a] depression of the spirit which has fallen upon so many of the people, making them, for the moment at least, listless, hopeless, without ambition." Caudill, who had deep roots in the area, considered the region "a colonial appendage" to the Northeast and Midwest.

Whatever success President Johnson's Great Society programs had in alleviating the suffering of the poor in the region, those programs couldn't help the decline of the coal industry. There were a few bright moments, such as the 1974 oil embargo, which caused coal prices to spike, thus creating a short boom. But then the long slide continued. More than half the mines in West Virginia closed by the 1990s. The ones that remained became bigger, more efficient, and less in need of human beings. Coal-rich McDowell County, which as late as 1980 had 7,200 mining jobs, the most of any of the eleven coalfield counties, had lost an incredible 90 percent of those jobs by 2003. Mercer County went from 890 jobs to 34. Overall, the southern coalfields lost about 26,000 mining jobs between 1980 and 2003. That's more than two out of three mining jobs in twenty-three years. What's left in the billion-dollar coalfields today is a lot of aging, sick, and dependent people, many of whom gave the best years of their lives in the mines and now find themselves poor and forgotten. In McDowell County, almost half of the income earned by the population is in the form of transfer payments such as welfare, workers' compensation, disability, Social Security, and other retirement benefits. The upswing in coal prices from 2003 to 2005 kept some marginal mines open that might otherwise have closed, but it has done little to change the overall economic picture in the region.

Although Appalachian coal operators like to blame environmentalists or overzealous federal regulators for the industry's decline, what's really doing them in is the state of Wyoming. Between 1997 and 2004, coal production in Wyoming grew 40 percent, while in West Virginia it dropped by 18 percent. Simply put, it is much easier and cheaper to mine coal in the wide-open, eighty-foot-thick seams of Wyoming than in the mountainous, mined-out, six-foot-thick seams of West Virginia. The difference in productivity between the two states is startling. In West Virginia, four tons of coal are mined per employee hour; in Wyoming, the rate is thirty-nine tons per employee hour — almost ten times more. In the past twenty years, the rise of the Powder River basin has had a devastating impact on Appalachia, putting tremendous pressure on operators such as Massey to slash costs or close down uncompetitive mines.

Wyoming is a stick in the eye to West Virginia in other ways, too. Like Appalachia, it's a fossil fuel colony, a place where the riches of the state are mined, drilled, and exported by big corporations that have only a passing interest in the welfare of the community. But Wyoming has learned from the mistakes of West Virginia. When the coal industry took off there in the 1970s, political leaders such as Republican governor Stan Hathaway had the foresight to understand what the hell-bent extraction of minerals and fossil fuels could do to a place. In 1974, Hathaway was instrumental in establishing the Permanent Mineral Trust Fund, which mandates that a percentage of all taxes collected from coal, oil, and gas companies go into a permanent fund for the welfare of future generations. Today the fund has $3 billion in it, spinning off about $100 million in interest each year, which is a good chunk of the state's $900 million budget. It has helped Wyoming build new schools and libraries, expand the state university, and improve recreation centers. The fund doesn't make Wyoming immune from the resource curse. Like West Virginia, it struggles with boom-bust economics and, more important, with the notion that ideas and creativity, not natural resources, are the foundations for prosperity. But the fund at least gives the state something to show for all the mining and drilling that has gone on, and it might help the state build a broader, more diverse economy down the road.

Ironically, the regions of West Virginia that are barren of coal offer the brightest economic hope today. The eastern panhandle of the state is attracting a number of high-tech firms, many of which are drawn there by its natural beauty and low cost of living. Morgantown and Huntington are emerging as hot spots for the new biometrics industry. Even so, despite repeated attempts by the West Virginia legislature to spur economic development, the shadow of the past has been difficult to shake. In 1987, for example, the legislature passed the West Virginia Capital Company Act, which was designed to give tax credits to companies as a way to encourage them to make venture capital available for new business start-ups. Ironically, two of the biggest recipients were coal companies: Riverton Coal Company received $1 million, and A. T. Massey (now Massey Energy) received $2.3 million. "That's venture capital in West Virginia — a scam to give a double dose of taxpayers' money to the coal industry while the economy languishes," wrote Denise Giardina, an outspoken West Virginia novelist and political activist who grew up in the coalfields. "This is corporate welfare fraud, pure and simple."

Down in McDowell County, nothing symbolized the region's tragic fall like a plan hatched in the early 1990s and recently revived to build a 50,000-ton-per-month megadump for garbage in the old coalfields. The dump would be roughly thirty times larger than the county needs for its own waste, the idea being to profit by importing garbage from nearby counties and states. Thus the sad transformation from billion-dollar coalfields to a million-dollar garbage dump would be accomplished. The company that first proposed this idea? The living ghost of Edward Berwind himself — the Berwind Corporation.

Not long after the Massey Energy Christmas Extravaganza, I visited Maria Gunnoe at her home in Bob White. Because she doubted I'd be able to find the place on my own, she met me in Madison, and I rode with her in a red Ford pickup for about twenty miles up into the steep hollow. More than once, she yanked the wheel to get out of the way of a loaded coal truck that swerved suspiciously close to the yellow centerline. "They all know my truck," she explained coolly. "Ever since I started talking in public about what the coal companies are doing, I've noticed that I'm not real popular with some folks around here." Not long ago, she discovered that her truck's brake lines had been slashed. Another day, her family's dog was found dead at her son's bus stop.

We turned onto a smaller road, approaching the Pond Fork River. Before we got there, the road was blocked with impromptu barriers. Beyond were the remains of a washed-out bridge that had been partially destroyed during the big flood in 2003 and was deemed unsafe for vehicular traffic. The West Virginia highway department had still not gotten around to fixing it, which meant that for nearly two years, Gunnoe has not been able to drive up to her house. Instead, she had to park beside the barriers and walk over the river on narrow wooden planks.

When I got out of the truck, Gunnoe pointed to a small blue and white house about one hundred yards away on the other side of the river. "That's my place up there," she said proudly. "I used to call it 'my little piece of heaven.'"

There was not much heavenly about it anymore. The house was on high ground, but it had the misfortune of being sited not far from where the Big Branch Creek tumbled out of the hollow to meet up with the Pond Fork. The hillside that the house was built on was rapidly being washed away. Besides wrecking the bridge, the flood had cut a gully about twenty feet deep and seventy feet wide in Gunnoe's front yard. It was a mean, nasty gash, still fresh-looking, as if someone had tried to cleave the hollow in two with a dull knife. Gunnoe explained that since the flood had wrecked the bridge, she had to lug everything — groceries, furniture, Christmas presents, even water — over the Pond Fork and up the hill. But that she could handle. What she could not handle was the peril of being washed out of her home.

We walked across the bridge and up toward her house. The Big Branch was just a trickle running out of the hollow. "The day of the big flood, we had ten feet of water roaring through here," Gunnoe explained. "It just came down in a fury. I thought we were goners."

Gunnoe's paternal grandfather, Martin Luther Gunnoe, a full-blooded Cherokee, labored underground in nearby coal mines for thirty-two years. He earned about eighteen dollars a week, and after many years of struggle, he was able to save enough money to buy these forty acres in Bob White. Gunnoe remembered helping her father and grandfather build the house. "I carried lumber, fetched the nails," she recalled. "We were proud of it. It was always real peaceful here."

The 2003 flood not only filled Gunnoe's barn with rocks and washed out her road; it also destroyed her septic system, ruined her well, and covered her garden with black sludge. Compared to others, however, she got off easy. All over the southern part of the state, homes were flooded, cars overturned, and lives destroyed. Seven people were killed, including a six-year-old girl who was drowned when the car she was riding in was washed off a narrow bridge. And that was after the 2002 floods, which had killed six people and destroyed more than two hundred homes. In 2003, West Virginia received $40 million in federal disaster relief aid, more than any other state in the country. No one blames the flooding entirely on the strip mines. In some areas, reckless timbering is also a factor. But poorly engineered mountaintop removal is the worst culprit. According to one EPA study, the runoff from these mine sites is sometimes three to five times higher than in undisturbed areas, which means that five inches of rain — about the amount that fell in southern West Virginia from June 13 to June 19, 2003 — has the same effect as would fifteen to twenty-five inches.

What bothered Gunnoe most was not losing her front yard to the floodwaters. It was the attitude of the engineer from the coal company who showed up the morning after. Gunnoe noticed him standing out in her yard and waded through the muck to see what he wanted. "The first thing out of his mouth was, 'We are not responsible for this. It was an act of God.'"

Gunnoe, who takes her faith seriously, and who had been up all night listening to the water roar and praying that she and her two children would not be swept away, was furious. Not only was she angry at the lack of sympathy his remark displayed, but she was upset by his passing the buck to God — who, in her opinion, would be outraged by what he saw the coal companies doing to West Virginia. "I didn't see God up there in those haul trucks, filling in the creeks," she shouted at him. "I didn't see God up there blasting the mountain with ANFO."

It took a while for the full tragedy of what had happened to sink in. But eventually Gunnoe realized that unless she could stop the mining, she would inevitably be forced to abandon her land. It was just a matter of time. This was reinforced a few months later when she was flooded again (not as severely). How long could she live in a place where every drop of rain struck terror in her heart? And it wasn't just the flooding. It was the fact that the place that was so much a part of her family life — where she knew every patch of ginseng and witch hazel, where she had spent the happiest moments of her childhood wandering in the woods — was being transformed into an industrial zone. The fish were long gone from the creeks, killed by the polluted runoff from the mines. The few black bears left in the area, aroused from hibernation by mining machinery, wandered around in an angry daze. The deer were gone, frightened off by the blasting. Soon she would be gone, too.

So she decided to fight. She quickly learned that no one at the county offices and none of her state representatives wanted to hear about her troubles. "They all just danced around my questions, or promised to call me back and never did," Gunnoe recalled. "It took me a while to figure out that they just didn't care. Or if they did call me back, they would talk to me about the importance of jobs. Jobs! When someone looks at me and says my job is more important than your life, they make an enemy of me."

Gunnoe began volunteering at Coal River Mountain Watch, a small group in nearby Whitesville that is headed by Julia Bonds, a former Pizza Hut waitress who has become one of the leading anti-mountaintop removal activists in the state. Gunnoe also attended rallies and public meetings with longtime activist Dianne Bady, the founder of the Ohio Valley Environmental Coalition, whom Gunnoe calls "one of the most courageous people I've ever met."

Bonds and Bady helped open Gunnoe's eyes to what was going on around her. She learned about the years of legal battles that environmentalists and local citizens had waged to slow mountaintop removal mining, and how the Bush administration had gone out of its way to subvert and delay those battles. Under the direction of Steven Griles, the deputy secretary for the U.S. Department of the Interior and a former coal industry lobbyist, debris from mountaintop removal mining was reclassified from objectionable "waste" to legally acceptable "fill," despite the fact that the debris is known to leach acid and heavy metals into local streams. This one-word change had a huge impact in Appalachia, undercutting the legal challenges to existing mountaintop removal mining and clearing the way for new mines. Gunnoe read one federal study that projected that over the next decade 2,200 square miles of land in Appalachia — an area larger than Rhode Island — would be impacted by mountaintop removal mining.

Gunnoe also learned about the dangers of coal slurry impoundments. In one instance, in 1972, the failure of a big slurry dam in Buffalo Creek, West Virginia, had sent a twenty-foot-high wall of coal slurry into the hollow below, killing 125 people and leaving 4,000 homeless. Today there are about 135 slurry impoundments in West Virginia, some of them the size of a good-size lake and holding billions of gallons of black water. One of the largest slurry dams in the state, the Goals impoundment in Raleigh County, is less than a mile above an elementary school. If the dam were to give out, the children wouldn't have a chance.

Gunnoe soon realized that flooding is only the most visible and melodramatic danger from coal slurry impoundments. Perhaps more threatening to the long-term sustainability of West Virginia is the leakage of these impoundments into the drinking water. One of the regions where the health effects have been of most concern is Mingo County, where Don Blankenship went to high school and still lives today.

A few years ago, Dr. Diane Shafer, a busy orthopedic surgeon in Williamson, the Mingo County seat, noticed that a surprising number of her patients in their fifties were afflicted with early-onset dementia. In addition, she was hearing more and more complaints about kidney stones, thyroid problems, and gastrointestinal problems such as bellyaches and diarrhea. Incidents of cancer and birth defects seemed to be rising, too. She had no formal studies to back her up, but she had been practicing medicine in the Williamson area for more than thirty years, and she knew that many people who lived in the hills beyond the reach of the municipal water supply had problems with their water: black water would sometimes pour out of their pipes, ruining their clothes and staining porcelain fixtures. Many people had to switch to plastic fixtures because steel ones would be eaten up in a year or two. The worst water problems were in the town of Rawl, near Massey's Sprouse Creek slurry impoundment pond, where millions of gallons of black, sludgy water is backed up. Were the health problems in the area related to the pollutants leaching into the water supply from the slurry pond? Dr. Shafer suspected they were.

Dr. Shafer is the lone physician on the Mingo County Board of Health. Despite her urgings, she could get no one at an official level to take much interest in the water problems in the area. So at her recommendation, a group of concerned citizens contacted Ben Stout, a well-known professor of biology at Wheeling Jesuit University and an expert on the impact of coal mining on Appalachian streams, to study the water quality in the area. Stout tested the water in fifteen local wells, most of them within a few miles of the Sprouse Creek impoundment and one just a short distance from Blankenship's home. Stout found that the wells were indeed contaminated with heavy metals, including lead, arsenic, beryllium, and selenium. In several cases, the levels exceeded federal drinking water standards by as much as 500 percent. Of the fifteen wells tested, only five met federal standards. Stout says that the metals found in the water samples were consistent with the metals in the slurry pond and the most logical explanation for how those metals got into the Williamson drinking water was that the impoundment pond was leaking into the aquifer. He also pointed out that coal companies often dispose of excess coal slurry by injecting it directly into abandoned underground mines, where it can easily migrate into the drinking water.

Dr. Shafer is the first to admit that much more work needs to be done to prove that the health effects she has been seeing are caused by the metals in the coal slurry. But state and federal agencies have shown little interest in pursuing it. "The coal companies control everything down here," Dr. Shafer told me. "It's like the Wild West — except there is no sheriff in town. They just do whatever they want and pretty much get away with it."

To Stout, the problems in Williamson are just a foreshadowing of what's to come in the rest of southern West Virginia. "We are taking one of the great fresh-water supplies in the world, and we're screwing it up," Stout says. "You can't fill in a thousand miles of streams, then inject millions of pounds of toxic slurry into the aquifers, without it having an impact." He believes that in twenty years, a lack of potable water may make southern West Virginia uninhabitable. "It's not the mountains of West Virginia that I worry about," he says. "It's the people. Sometimes I think what's going on here is damn near genocide."

The question is not whether coal mining as it is currently practiced in West Virginia is sustainable. It clearly is not. The question is whether the modest economic benefits of another decade or so of mining are worth the environmental and human damage that will go with it.

Despite the public rhetoric about jobs that accompanies any debate about coal mining in West Virginia, the illusion that coal will bring wealth to anyone but Don Blankenship and his kind faded years ago. Even Senator Robert Byrd, a champion of coal for more than fifty years, has acknowledged that West Virginia needs to start thinking differently about its future. But after more than one hundred years of betting on the wrong horse, West Virginians find themselves in a tough spot: how does the state begin to imagine a future without coal? West Virginia is hooked on coal tax revenues. Even a modest decline would be a huge blow to the state budget, forcing cutbacks in services that would be political suicide for any elected official who presided over them. When the economics of the state worsen, so does its dependence on coal. And that, it turns out, is very good news for companies like Massey Energy.

Consider this TV commercial that Massey ran in West Virginia in

2004, around the time it was pressuring the Army Corps of Engineers to speed the permitting process on its mountaintop removal mines. The ad opens with a young mother and father sitting down to have a conference with their son Johnny's teacher at an elementary school. The teacher tells them that Johnny is just not himself in the classroom these days — is something going on at home? Anguished looks are exchanged between the mother and father. Well, yes, the mother blurts out. The coal mine where Johnny's dad has worked for eighteen years is shutting down. Now they're going to have to move out of the state to find work, and Johnny is very upset about it. Why is the mine shutting down? "We've mined out all the coal we have a permit for," the father explains angrily. "The permits for additional areas have been held up for four years." The teacher shakes her head, and a sorrowful look appears on her face. "I've seen so many families in the same situation," she says.

The screen fades to black, and a graphic appears: "Since 1990, West Virginia has lost about twenty students per school day — a loss of 38,000 students." A voice-over says: "Support our kids, their schools, and their teachers by supporting coal. Coal means West Virginia jobs at home for West Virginians."

This ad is just one in a series produced by Massey, but all the ads have a similar theme and message: West Virginia is in trouble, and it's because bureaucrats, regulators, union organizers, and environmentalists are getting in the way of mining coal. Those groups are never mentioned by name, of course, but the implication is clear to any viewer. This argument is astonishing on many levels, not least because it so baldly holds that Big Coal is the *solution* to the state's woes, not the cause of them. If there is any state in the Union that symbolizes what more than one hundred years of unchecked, unrestrained coal mining can do to a place, it's West Virginia. The 13 billion tons of coal that have already been dug out of the state have brought little more than heartache and poverty. *But if you will just let us blast one more mountain, haul out one more trainload of coal, everything will be okay.*

The irony — and genius — of this argument is that the farther the state's economy declines, the more potent the argument becomes. It's one thing to lose coal mining jobs in a growing, diverse economy; it's quite another to lose them when it's the only game in town. Don Blankenship has apparently figured out that if you keep people barefoot and pregnant, they'll work cheap and love you more. As Kris Warner, the chairman of the Republican Party in West Virginia, put it during the 2004 presidential campaign, "If the likes of Massey Energy closes their doors in West Virginia, there will be absolutely no hope [in the southern part of the state]."

Massey, of course, has often been willing to bust unions, cut benefits, or lay off workers if doing so meant lowering the cost of business. But if a permit is stalled, it's the tree-huggers and Washington bureaucrats who are turning the lights out in West Virginia. And Massey is hardly the only coal operator guilty of such tactics. When Arch Coal, the St. Louis–based corporation that is the second-largest coal producer in America, was forced to idle its huge Dal-Tex strip mine in West Virginia because the Clean Water Act forbade the company from filling in any more streams, Arch laid off 350 workers and blamed it on overzealous regulators. But an editorial in the *Charleston Gazette* pointed out the hypocrisy of Arch's claim: "This company has never hesitated to lay off workers, especially union workers. But when asked simply to follow a twenty-two-year-old federal law, the company treats the loss of 350 jobs as a traumatic event. Arch Coal is using those 350 jobs to try to blackmail regulators and government officials into ignoring the law. If Arch Coal were successful and the permit for the expansion of Dal-Tex were approved today, does anyone really believe that Arch Coal would hesitate to fire all those miners in a year or two for its own reasons?"

If there was any doubt that Blankenship is the biggest bully in the sandbox, those doubts were vanquished during the 2004 election, when he dumped $3.5 million of his own money into a campaign to defeat state Democratic supreme court justice Warren McGraw. Blankenship's money was funneled into a 527 group, one of the supposedly independent political organizations that were created by a loophole in campaign finance reform laws. (The organization Swift Boat Veterans for Truth, which played such a prominent role in the 2004 presidential election by discrediting Senator John Kerry, was a similar 527 group.) Blankenship's group was called And for the Sake of the Kids. It funded a high-profile ad campaign criticizing

McGraw for signing a state supreme court decision to parole a convicted child molester and allow him to work as a janitor in a public high school. The child molester, who had pled guilty to first-degree sexual assault when he was fifteen and had been a victim of child abuse himself, was hardly your stereotypical sexual predator. Nor had he actually been released to work in the high school; after the decision was signed, authorities had thought better of it. The mere fact that McGraw, along with two other justices, had signed the decision was enough to allow And for the Sake of the Kids to run a series of widely publicized campaign ads suggesting that McGraw coddled child molesters.

The real reason for Blankenship's interest in the race may have been less sensational. In 2002, a Boone County court had found Massey guilty of fraud in its negotiations to buy a rival coal operator and had leveled a $50 million judgment against the company. Blankenship made no secret of the fact that he intended to appeal the judgment to the state supreme court. McGraw was known to be no friend of the coal industry, and with the court evenly split, Blankenship was eager to see him replaced by a more pro-business justice. Blankenship's candidate of choice was Brent Benjamin, a Republican lawyer who had spent much of his career in private practice defending corporate interests and could be counted on to be much more sympathetic to the arguments of Blankenship's lawyers. In short, by spending $3.5 million or so to unseat McGraw, Blankenship was likely to reap $50 million in rewards.

Benjamin won 53 percent of the vote. The day after the election, in response to a reporter's question, Benjamin would not promise to recuse himself if the case against Massey was heard on appeal, insisting with a straight face, "I'm not bought by anybody." One of Benjamin's fellow supreme court justices, Democrat Larry Starcher, had a different opinion. A few weeks later, he marked Benjamin's arrival at the court by taping an editorial cartoon on his office door. It depicted Benjamin showing up at the high court in a shipping crate marked "courtesy of $3.5 million from Massey Coal and other coal interests."

By 2006, West Virginia voters had wised up to Blankenship's cynical tactics. The Massey CEO spent over $2 million during the midterm elections, doling out cash to his favorite Republican candidates for the state legislature and turning the Grand Old Party into just another Massey subsidiary. But of the forty or so Democrats on Blankenship's hit list, only one failed to win re-election. As the *Charleston Gazette* put it in a morning-after editorial: "The poor showing of almost all Republican nominees backed by Massey Energy CEO Don Blankenship affirms an ageless truism: Money can't always buy everything."

Sitting on the couch in her small, neatly kept house, Maria Gunnoe changed into her hiking boots, then disappeared into another room for a moment. When she returned, she was carrying a silver .32-caliber Colt pistol. "It was my grandfather's," she explained. "In case we run into bears." Or angry coal miners? "Unlikely," Gunnoe said matter-of-factly. "But you can't be too careful up there."

Gunnoe tucked the pistol under her belt, and we headed up the hollow, following the steep cut of the Big Branch. You could see more evidence of the floods — huge cracks and fissures cut in the earth by the rushing water, trees upended, their roots twisting up toward the sky. It was muddy and slick in places, but Gunnoe climbed with ease. She talked about how spring is the scariest time of year for her now. How big will the flood be this year? How fast will it come? She knows parents in Bob White who make their kids sleep in their clothes when it rains so that they're ready to go when the waters hit.

As we hiked, I asked her the obvious question: why not leave?

"If I leave, where am I going to go? This is my place, it is who I am. My memories are here, my life is here. I love this place. Who am I if I give all this up? *Why should I give all this up?* To make it easier for the coal company? Should I just admit that Don Blankenship has won, let him take over the state? My friends tell me I'm crazy, that this place is already ruined, so what's the point? But I'm not one to give in. My feeling is, you fight for what you've got, even if it's only worth a dime."

A half-hour or so later, the hollow broadened. We passed several NO TRESPASSING signs that marked the mine border. To Gunnoe, these signs meant less than nothing. We walked through a stand of pines and poplars, then abruptly confronted a huge wall of rock — a valley fill.

"Ain't it pretty?" Gunnoe said wickedly.

I had seen many valley fills before, but never quite from this angle. It was a wall of rock maybe five hundred feet high, barren of trees, but with thin patches of grass growing here and there. Standing at the bottom, looking up, I thought it seemed both immense and fragile. It was, quite literally, the top of a nearby mountain that had been cut off and dumped in this narrow valley that had once been the headwaters of the Big Branch. Culverts had been installed to divert the water, but in the end there was nowhere for it to go but down the hollow. In the distance, we could hear the faint grind and roar of heavy machinery. Every few minutes, a haul truck would appear on the horizon at the top of the mountain, hauling a load of dirt and rock to another part of the mine. On one of the steps in the valley fill, the company had planted several fruit trees and carefully surrounded them with chicken wire to keep rabbits from gnawing at the bark. It was a touching gesture, but one that somehow made the larger devastation all around us seem all the more criminal. Gunnoe pointed out that one of the trees had already been smashed by flyrock from a mine blast.

Gunnoe makes this hike up to the mines several times a week. For her, it is a way of claiming ownership of the land, and of reminding the coal company that it's being watched. It's a risky trek — once she was almost hit by flyrock; another time she was confronted by a group of miners who made vague remarks about how dangerous it was for a woman to be walking alone in the woods. "The next day, I hiked up here with my shotgun and said, 'Who's first?' They left me alone after that," Gunnoe said.

Seeing Gunnoe standing in front of the valley fill reminded me of the famous picture of the student standing in front of the Chinese tank during the Tiananmen Square uprising. In a way, this valley fill is Gunnoe's tank. Her life, her feelings for her land, her memories of exploring the woods as a kid and of helping her grandfather build his house — all that means less than nothing in the face of the industrial operation above us. She doesn't care. "They can bury me in these hills," Gunnoe said, facing the wall of rock, "but I ain't leavin'."

Chapter 3

Dogholes

PROMOTING "CLEAN COAL" has often meant keeping the public in the dark about coal mining. Ads by power companies and coal industry trade groups usually feature panoramas of bright city lights or images of kids playing baseball on a well-lit baseball diamond, not black-faced miners coughing up soot. This is hardly surprising. The gritty reality of coal mining — especially in underground mines — is not one of coal's competitive advantages: roughnecks working on gas rigs may lose fingers with unfortunate frequency, and workers installing solar panels may risk tumbling off the roof, but coal is the only energy source that requires workers to put their lives on the line on a daily basis. Would Americans be quite so eager to burn coal if they knew what life is like in a coal mine? Maybe not.

Oddly enough, the long-held industry taboo against spotlighting the dirty reality of coal mining was recently broached at a glitzy cocktail party on Pennsylvania Avenue in Washington, D.C. General Electric threw the bash in May 2005 to celebrate the launch of "ecomagination," the company's high-profile initiative to ramp up development of clean energy technologies. Jeffrey Immelt, GE's chairman and CEO, mingled with congressional representatives, power industry executives, environmentalists, and James Connaughton, President Bush's senior environmental adviser. Invitees sipped wine produced by a solar-powered California vineyard (equipped with GE's own photovoltaic panels) and perused exhibitions of the company's new technologies — a life-size model of a hybrid-engine train, a state-of-the-art wind-turbine blade, a superefficient washing machine, and even a diorama of coal gasification technology.

In a brief speech, Immelt announced that "it's no longer a zero-sum game — things that are good for the environment are also good for business." He vowed that GE was embarking on this initiative "not because it is trendy or moral, but because it will accelerate [economic] growth." He then presented a series of "ecomagination" television ads that would soon be airing nationwide.

One spot in particular, which Immelt described as "a play on how to make coal sexy again," elicited applause from the guests. As most people in the room knew, making coal sexy is an important goal for GE, which is a major supplier of high-tech turbines and other power plant technologies. The ad featured glamorous, scantily clad models (male and female) with chiseled pecs and perky breasts, shoveling coal in a dark mine — actually, a sound stage outfitted to look like a mine — while "Sixteen Tons," Merle Travis's great folk song about hard labor and corporate exploitation, played on the soundtrack. Near the end of the ad, a voice-over announced, "Now, thanks to emissions-reducing technologies from GE, harnessing the power of coal is looking more beautiful every day."

It was a bold ad, and one that suggested how well GE understands a basic truth about coal: Americans love cheap power, but not if we have to pay for it with the kind of human misery and suffering that went out of style in the nineteenth century. However, if the point of GE's ad was to clean up the image of coal mining to shore up support for "clean" coal, it backfired. "The commercial is sure to disgust and anger anyone who grew up in a mining community," one viewer wrote in a letter to the *Chicago Sun-Times* a few days after the ad was broadcast. "Even if coal processing gets cleaner, that coal will still need to be mined," wrote ad critic Seth Stevenson on *Slate*, an influential online magazine. "And unless I'm mistaken, there will be actual coal miners doing that. Now: Guess who still gets black lung? Guess who still gets killed when mines collapse? It isn't sexy supermodels."

It's people like Randy Fogle, one of the nine miners whose dramatic rescue after being trapped for seventy-seven hours underground near Quecreek, Pennsylvania, captivated the nation in 2002. Fogle is about as far from a supermodel as you can get and still be considered a member of the same species. He is in his early forties, a big, bearlike man with blue eyes, a mustache, and a thick wad of chew under his lower lip. When I first met him a few weeks after the rescue in his small, dark house in the hills near Quecreek — I had been hired to edit a book by the miners about their experience — he seemed married to his La-Z-Boy chair. When he got up, he moved gracelessly, with a limp. I had a difficult time imagining how he could maneuver in the tight confines of an underground mine.

For Fogle, the romance of coal mining isn't just a marketing ploy. He visited his first mine, which was owned by his grandfather, when he was six years old. He still remembers the mystery of it — the damp, earthy smell, the urgency of the work, the bond between the men. There was never any question about what Fogle would do with his life. He went underground after high school and never came out. During his twenty-two years in the mines, Fogle had known dozens of men who had been killed or maimed in the darkness below (including his grandfather and his wife's grandfather). He had seen the unions run out of the area and had experienced firsthand the falling wages and the cuts in health care benefits that resulted. He knew all about thinning coal seams and greedy mine bosses and the dangers of black lung. But none of that distracted him from his desire to dig coal.

When Fogle talked about life underground, his whole demeanor changed: he leaned forward in his chair; his movements quickened; he talked faster. He described the clarity that comes with the knowledge that one wrong move could send thousands of tons of rock down upon his head, and the odd exhilaration he feels going into dark places in the earth where no man has gone before. "It's Mother Nature you're playing a game with; it's just you and her," Fogle said, glowing with excitement. "You're playing against a force that's awesome."

A few weeks after we met, I accompanied Fogle when he went underground for the first time since he had been pulled out of the flooded mine at Quecreek. It was a working mine a few miles from his house that was nearly identical to the one he'd been trapped in

— deep, low, and rough. In the changing room, we donned coveralls and mining boots, as well as leather belts with breathing devices strapped to them and numbered brass tags that could be used to identify our bodies if necessary. Fogle was cool, serious, but obviously eager to get underground. We climbed onto the mancart — a low golf cart–like vehicle that is commonly used to carry workers into the mine — leaned back, and began our descent. It was a twenty-minute trip through absolute darkness, filled with weird rumblings and groanings of the earth. Finally, we arrived at the coal face, which was lit by a few small lights. It smelled earthy and wet, reminding me of my wife's garden after a thunderstorm. The mine was about four feet high — too low to allow us to walk. The only way to move around was to crawl, or half crouch, half squat in a position miners call a duck walk.

Underground, Fogle was a different person. Despite the difficult conditions, a kind of grace and elegance overtook him. He moved like a big cat, using his pick as a walking stick as he scrambled around in the darkness, moving twice the distance in half the time as I could. He seemed to anticipate the movement of the machinery around him, always in the perfect spot to lend a hand to one of the working miners. I tried to follow his eyes, to see what he was seeing, but he was attuned to the darkness in ways that I could not understand. This is a man, I realized, who was born to mine coal. It is a fate that had already nearly killed him but that has also brought him vividly to life. "Coal mining is in your heart, it is in your blood," Fogle told me later. "Otherwise, you might as well forget it."

The rescue of the Quecreek miners did for coal mining what GE's "model miners" ad couldn't: it gave miners glory, acclaim, and respect in the eyes of millions of Americans. The rescue was a perfect made-for-TV moment, in part because it bore a striking resemblance to the scene at the World Trade Center that had so horrified us a few months earlier. Except this story had a happy ending — all nine men rose out of the earth, spiritlike, covered with soot, seemingly no worse for the wear. For a few weeks, these nine men were the talk of the nation. They were flown in private jets to NASCAR races, appeared on *The Oprah Winfrey Show*, and were justly celebrated by President Bush for representing "what I call the spirit of America, the great strength of our nation."

Then it was over, and their names and faces were quickly forgotten. As for the tough questions about why these men had almost died, or what this near-tragedy said about the business of mining coal in the twenty-first century — well, that would presumably be answered by the half-dozen agencies and blue-ribbon panels that were looking into what had happened. But from the beginning, responsibility was a sideshow. The main event was the miracle.

As it turned out, there was much to learn about the coal industry, particularly about how it treats its workers, from what happened at Quecreek. If things had gone a little differently — as it did, tragically, for the twelve men in the Sago mine near Tallmansville, West Virginia, in January 2006— there might have been a little less focus on divine intervention and a little more focus on the nuts and bolts of why those men had tunneled anywhere near an abandoned mine that everyone knew was flooded with billions of gallons of water. Not that a celebration wasn't justified. The drillers and engineers who orchestrated the rescue deserved all the credit they got for their extraordinary work of getting those men out of the mine alive. But their success stood in stark contrast to the recklessness and negligence that allowed the men to become trapped in the first place.

Big Coal works overtime to suggest that coal mining today has nothing in common with its dark and exploitative past. The old days of breaker boys and methane explosions and black lung are gone, the coal industry argues, and mining today is safe, well paid, and professional. Images of coal miners in promotional literature published by the National Mining Association and others always show clean-faced men in close proximity to high-tech machinery — computer screens, Global Positioning System (GPS) equipment, bright yellow haul trucks. The average salary of a coal miner, the association boasts, is $50,000 a year. According to one coal industry Web site, working in a coal mine today is as safe as working in a grocery store.

Of course, that is nonsense. Working in a coal mine in West Virginia may be much safer than working in a coal mine in China, but according to the National Institute for Occupational Safety and

Health, mining is still one of the most dangerous occupations in America. Working in an underground coal mine, especially one owned by a small, non-union operator, such as Quecreek, is most dangerous of all. The fatality rate in these mines is five times higher than it is in surface coal mines. Although mining fatalities fell precipitously after the passage of the Federal Coal Mine Health and Safety Act in 1969, the numbers have leveled off since then. Many of the deaths in recent years could have been prevented. In one case, after thirteen men were killed in a methane explosion in an Alabama coal mine in 2001, investigations indicated that the mine openly flouted basic safety regulations. Similarly, after the Sago mine exploded in West Virginia in 2006, it was discovered that the mine had been cited for more than two hundred federal safety violations during the previous year. And despite the coal industry's claims that black lung is a disease of the past, more than 1,500 (mostly retired) miners still die from it every year.

Claims about the high wages earned by today's coal miners are also distorted. Like most blue-collar workers, coal miners have not shared in the bounty of economic progress during the past several decades. According to U.S. Department of Labor statistics, the average weekly wage of a coal miner in 2004 was about 20 percent lower than it was in 1985 (adjusted for inflation), despite the fact that the average miner's per-shift productivity had nearly tripled in the same period. Wages vary widely from mine to mine. In the East, union mines pay 20 to 30 percent more than non-union, and nationwide strip mines generally pay better than underground mines. A heavy equipment operator at a big strip mine in the West can make $60,000 a year or more. By contrast, an underground laborer in a mine like Quecreek, where the risk of injury and death is far higher, makes only half that. Although coal jobs are often touted as work that can't be outsourced to China, the industry has consistently threatened to shut down mines and cut back operations in order to force workers to agree to wage and benefit cuts. Thanks to aggressive union busters like Don Blankenship, as well as years of misguided union leadership, union mines in the East have been hardest hit. The United Mine Workers of America, once one of the most powerful voices for workers in the country, has seen its membership decimated in recent decades. Today the union represents only about 40 percent of U.S. mineworkers, and the percentage is falling every year.

Unlike Randy Fogle, Blaine Mayhugh, the youngest of the nine miners trapped at Quecreek, never dreamed he'd work underground. He grew up in Meyersdale, a small town in the heart of Pennsylvania's old bituminous coalfields, but his father worked in construction, and he always imagined he'd do the same. During high school, Mayhugh was a prankster with a good fastball who preferred hunting and fishing to reading and writing. After graduation, he joined the navy and thought about making a career of it, but he was discharged after a fight in a New Jersey bar. When he returned to Somerset County, he married Leslie Foy, his high school sweetheart, and went to work at a metal fabrication shop. Then he took a job with a chemical company, spraying herbicide on lawns for eight dollars an hour.

Like many old coal regions, Somerset County never developed the traditional steppingstones to help guys like Mayhugh build a life. It is one of the few counties in Pennsylvania with no community or state college and no vocational schools to speak of. The resources of the big city, including the University of Pittsburgh, which has helped thousands of laid-off steelworkers reimagine their futures, are an hour and a half's drive away over a nasty turnpike. The city may as well be on another planet. What few manufacturing jobs there once were in Somerset County vanished long ago. There are really only three career options to speak of in the area: the state prison, the coal mines, and Wal-Mart.

It was a different situation when Mayhugh's father-in-law, Tom Foy, returned home from Vietnam in the 1970s. At that time, thanks to the oil embargo in the Middle East and the sudden awareness of the economic and political risks of our dependence on foreign energy, coal was back in fashion, selling for seventy dollars a ton or more. Mines were opening up all over the county, and engineers and surveyors were in high demand. Learjets buzzed investors in and out of the Somerset County Airport; women in furs and diamond tiaras appeared at the Ramada Inn. Foy followed the money and went underground. For a few years, he earned a decent paycheck, including bonuses when his crew outloaded the other crews in the mines. Then, just as quickly as it had started, the boom ended. Prices fell, mines shut down, and workers were laid off. Foy was one of the tough ones. He held on, moved from mine to mine, and always found a way to make a living.

After Mayhugh and his wife had their second child, it was clear that he couldn't go on spraying lawns forever. Leslie, who worked part-time as a nurse's aide in a local retirement home, suggested he talk to her father about getting a job in the mines — not forever, but just for a few years, while the kids were young. Mayhugh had heard plenty about life underground from Leslie's father and knew the problems he had with his knees and his back from spending ten hours a day crouched in a dark hole. Mayhugh knew it was not a good job to grow old in. Living in Somerset County, he had seen plenty of old men on respirators — "puffers," they called them — whose lungs were stiffened by years of inhaling coal dust. Mayhugh hoped for something better. He applied for work with the turnpike authority and the state prison, but when those attempts failed, he didn't know where else to turn. It took Foy a few months of lobbying to get Mayhugh a job, but finally there was an opening. Mayhugh was hired by the same company Foy worked for, PBS Coals, one of the largest operators in the region. Mayhugh went through a brief training program, then was put to work underground in a mine called Barbara (which, it would later turn out, was directly beneath the open field where United Airlines Flight 93 crashed on September 11, 2001). Barbara did not live up to its gentle name: the mine was low, wet, and rough — what miners call a "doghole."

A doghole today is not like a doghole a hundred years ago. Back then, there were no labor laws, no safety helmets, no mine inspectors, no safety training, no high-tech ventilation fans, no heavy steel roof bolts, no steel-toed boots, no methane meters to detect dangerous levels of the invisible, explosive gas that often is released in coal mines. One hundred years ago, coal mines were like war zones. Edgar Allen Forbes, an essayist and activist, described the scene in Monongah, West Virginia, after a methane explosion in 1907 killed 361 workers: "Think of hell as a hollow hill and imagine that its power plant has exploded and blown a hole in the hillside. Then imagine a handful of reckless, begrimed men going into the cavern with lanterns, with sulfurous fumes in their faces, and dragging out the charred bodies of men . . . That is what Monongah looked like." (Some historians believe that as many as 550 miners were killed at Monongah, but we'll never know for sure. Although the coal company kept an accurate record of the number of mules working in the mine, it never bothered to count the number of men, and after the disaster, the mine was sealed up before all the bodies could be recovered.)

It's difficult to imagine anything like Monongah happening today. But even one hundred years' worth of safety innovations and better equipment have not changed the fact that work underground, especially in the low, tight mines of Somerset County, is still a dangerous and brutal business. "From the first day I went underground," Mayhugh told me later, "I always knew there was a chance I would die in there."

Although PBS Coals, which owned both the Barbara and Quecreek mines, had a local office nearby, it was hardly a local company. PBS was a subsidiary of a British holding company called Mincorp, which in turn was largely owned by Citigroup and the Bank of Scotland. The company was run by a dapper, clean-shaven British businessman named Robert Scott (the miners' wives referred to him as "Mr. Penny Loafers" because he often showed up at the mine site in soft brown loafers) and was typical of the low-end coal companies that thrived in the area after the 1970s boom went bust. It was a strictly cut-rate, non-union operation, with a reputation among locals for using shoddy equipment and squeezing every last dime out of every ton of coal. PBS specialized in exploiting old, hard-to-reach coal seams that were too complex or dangerous for other coal companies to bother with. They sold most of their coal to the nearby Homer City and Conemaugh generating stations, both of which are among the most highly polluting power plants in the region, and were aided by federal synfuel tax credits, a much-abused federal subsidy that gives coal operators big tax breaks for "processing" coal in various ways to make it burn cleaner. (The environmental benefits, if there are any, are slight.) For liability reasons, the company rarely ran the actual mine operations and instead subcontracted the work out to independent operators who paid PBS a percentage of every ton of coal removed from the ground. Such an arrangement had many advantages for PBS, not the least of which was that it allowed them to close down any mine at which the workers got too uppity.

Mayhugh, of course, knew nothing about this when he first went underground. He just wanted to survive. He was bigger than most miners, about six feet tall, and crawling around in a dark, wet hole about forty inches high for ten hours a day was a shock to his body. He was sore for weeks. Like every new coal miner, he was hazed by the older guys and given all the toughest work. But Mayhugh won them over with hard work and practical jokes. He'd defecate at the coal face and leave it for one of the older guys to discover, or sneak into their lockers and fill their helmets with talcum powder. Mayhugh liked the camaraderie underground, especially the endless talk about hunting, fishing, and sex — Mayhugh's three favorite subjects. He also liked the paycheck. Thirteen dollars an hour felt like a king's ransom to him. It allowed Leslie and him to put a down payment on a little place in Meyersdale that backed up to a creek and was only a block away from Mayhugh's parents.

Still, Mayhugh admitted he was scared. One day, when he was driving a shuttle car, he pulled up behind the mining machine — a big, carbide-toothed buzz saw that cuts the coal — and a rock the size of a softball went shooting by his head. He worked with guys who had seen their friends killed, who had been buried under rockfalls or badly injured by mining equipment. Like most new coal miners, he tried to keep his mind off that stuff and just focus on loading coal, but it was nearly impossible: in a mine like Barbara, near-misses were an almost weekly occurrence. Eventually, the conditions got so bad that production plummeted and the mine was shut down. Mayhugh found himself transferred to another mine, called Sarah, for a few months. When they ran into trouble there, he was moved again, this time to Quecreek.

Quecreek was another PBS scavenger operation. Coal seams in the Quecreek area had been mined on and off for a century, and the rolling hills were literally hollowed out with tunnels. But in the late

1990s, PBS was able to cobble together ownership of a modest tract of coal that had not yet been touched. It would be a risky operation, in part because to access the coal, the company would have to cut a new opening in the hillside, which was always an expensive proposition. What if workers mined in one hundred feet and hit a granite ledge? PBS could easily invest several million dollars in the mine, then find that it was all but worthless.

The other problem was that there were a number of old mines in the area, including one known as the Saxman mine, which, like many old mines, was flooded with billions of gallons of water. This was hardly a secret. In 1999, when the Pennsylvania Department of Environmental Protection (DEP) held public hearings before granting the permit to PBS for the Quecreek mine, dozens of local residents warned the company about the old mine. Everyone knew it was flooded — some residents were even using it as a well for their drinking water. Besides the obvious danger of running into the old mine, residents worried that the new mine might pollute or draw down the aquifer in the area, causing problems with the drinking water. For these and other reasons, it was vital that the exact boundaries of the old mine be carefully established. But it was clear even to local residents like Jeffrey Bender, who filed one of forty-one complaints against the proposed mine in 1999, that something was amiss in the permitting process. In a letter sent to the DEP at the time, Bender wrote, "Obviously the firm preparing to mine hasn't researched the area very well at all." The DEP, which had jurisdiction over mining permits, assured residents that no new mining would be allowed until inspectors were certain it was safe.

Abandoned mines are not a problem unique to Somerset County. By one count, Pennsylvania is honeycombed with more than 1,700 old mines (engineers often joke that the entire state is held up by old locust posts). West Virginia and Kentucky are nearly as bad. Each year, half a dozen or so people in these states die by falling into old mines. And that's just part of the problem. Coal fires burn in abandoned mines for decades, spewing dangerous levels of carbon monoxide. Mines silently flood for years before suddenly disgorging millions of gallons of red, acid-laced water, poisoning nearby streams and rivers. Sometimes the ground over mineshafts collapses, opening up huge holes. They undermine the foundations of nearby roads and buildings. Pennsylvania officials estimate it will take sixty years and cost fifteen billion dollars to clean up and seal the old mines.

For underground coal miners, abandoned mines are a deadly hazard. More than half the underground mines in Pennsylvania are adjacent to abandoned mines, and every coal miner who has spent some time underground knows that hitting a flooded mine is very bad news. In addition, experienced coal operators understand that the precise location of an old mine is notoriously difficult to determine, not only because old maps are often unreliable but also because coal companies in the past often "stole" coal by mining more than they were permitted to take. In 1999, the DEP clarified a rule stating that any mining operation that came within one thousand feet of an abandoned mine must have the borders of the old mine verified, either by producing a map stamped "final" by an accredited engineer or by conducting a new survey. Furthermore, the rule stated that if a mining operation came within two hundred feet of an old mine, boreholes had to be drilled to determine the precise location of the old workings. Safety-conscious mine operators often drill ahead more frequently than is required by law, especially in areas like this, where a flooded mine was known to be nearby. It costs time and money to survey the old mines, but what's a few thousand bucks if it means guaranteeing the safety of the workers, not to mention protecting your investment in the mine itself?

PBS Coals, along with a consulting firm hired by the company, spent a year looking in various state and local repositories for a final map of the Saxman mine, but apparently they never came up with one. Instead, they found half a dozen other old maps, the latest of which had been drawn up in 1961, around the time the Saxman mine had supposedly quit operations. It was not marked "final," however. At this point, a prudent company might have done further research — either by drilling test holes or using two-dimensional seismography — to determine the exact boundaries of the old mine. PBS did neither. Instead, the company showed the DEP the old maps they had found and said, in so many words, This is the best we can do. And the DEP, ignoring its own rules, community concerns, and common industry practice, agreed that it was good enough. In 2001, the permit for the Quecreek mine was granted, and mining operations began soon after.

Mayhugh didn't arrive at Quecreek until it had been in operation for several months. As soon as he began work, he heard talk about a flooded mine somewhere in the area, but he never paid much attention to it. He had enough to worry about, becoming familiar with the new mine and surviving each shift without getting crushed by machinery. Mayhugh had been working for PBS long enough to know that the miners' safety was not always at the top of the company's agenda. But like every other coal miner in the place, he assumed that the guys who ran the mine or signed off on the permit had checked things out. A coal miner is not paid to ask questions; he is paid to load coal.

If the sorry history of the coal mining industry has proven one thing, it's that when it comes to enacting and enforcing safety laws against Big Coal, the only good lobbyists are dead miners. Even the public outrage that followed the disaster in Monongah in 1907, which left 250 widows and 1,000 fatherless kids, was not enough to push Congress to pass safety laws against the all-powerful coal industry. It took another dozen mine disasters throughout the country, and another 1,200 or so dead miners, before Congress finally acted, creating the U.S. Bureau of Mines in the Department of the Interior in 1910, with instructions to investigate mining methods, "especially with respect to miners, and . . . the possible improvements of conditions under which mining operations are carried on." But the legislation provided no enforcement power whatsoever, and investigators could enter a mine only with the permission of the owner and were not allowed to publicize their findings. It took another three decades — and many thousands of dead coal miners — before Congress granted the bureau the authority to inspect mines and publicize its findings. But it still had no enforcement power.

In 1947, an explosion in a mine in Centralia, Illinois, killed 111 miners. An investigation later revealed that years of warnings about dangerous conditions in the mine were contemptuously ignored by the mine owner. Testifying before Congress, John L. Lewis, the powerful head of the United Mine Workers of America, thundered, "If we must grind up human flesh and bone in the industrial machine we call modern America, then before God I assert that those who consume coal and you and I who benefit from that service because we live in comfort, we owe protection to those men first, and we owe security to their families if they die." In 1952, President Harry Truman — one of the few U.S. presidents who had the courage to stand up to Big Coal — signed the Federal Coal Mine Safety Act. The legislation was full of loopholes, written into the law by coal industry lobbyists, but for the first time, it gave mine inspectors the power to shut down certain types of dangerous mines.

It still didn't stop the deaths. In 1968, an explosion in Consolidation Coal's No. 9 mine in Farmington, West Virginia, killed seventy-eight miners. The mine had had a history of accidents. It had blown up in 1954, killing sixteen men, and in the two years before the 1968 explosion, inspectors had cited Consolidated Coal for numerous safety violations. The mine was destroyed by the 1968 blast and burned for several days. To smother the flames, the mine was sealed shut, and the bodies of the miners were never recovered. Once again, there were calls for the government to crack down on Big Coal. "Let me assure you," Secretary of the Interior Stewart Udall told a conference on mine safety, "the people of this country will no longer accept the disgraceful health and safety record that has characterized this major industry." The following year, Congress passed the Federal Coal Mine Health and Safety Act, which dramatically increased the enforcement powers of the Bureau of Mines. It also gave miners the right to request a federal inspection and, for the first time, provided benefits to miners disabled by black lung. It was landmark legislation, but it had come too late for the nearly 100,000 coal miners who had already been killed since 1900.

In the years after the bill was passed, the rate of fatal accidents declined gradually and irregularly. President Richard Nixon helped stall any tough enforcement by installing political cronies in top positions at the bureau. In 1972, the General Accounting Office reported that "the [Interior] Department's policies for enforcing health and safety standards have been extremely lenient, confusing, and inequitable." In part because of this, in 1977 the power to inspect mines and enforce safety laws was transferred from the Department of the Interior to the Department of Labor, where a new agency, the Mine Safety and Health Administration (MSHA), was created. But problems of failing to enforce the law against coal companies continued. In 1987, Republican senator Orrin Hatch called MSHA "an agency in trouble," with a "disturbing pattern of misconduct, mismanagement, and serious abuse."

During the Clinton administration, MSHA was presided over by J. Davitt McAteer, a lawyer and mine safety expert who was no friend of Big Coal. Then, in 2001, President George W. Bush restored the tradition of cronyism and lax enforcement by giving the top job at MSHA to Dave Lauriski, a former executive with Energy West Mining Company, a small underground coal producer based in Utah. Although Energy West's mines had a good safety record, Lauriski was known for his belief that the coal industry should be given increasing latitude to police itself. He had spent years lobbying MSHA to loosen the rules against dangerous levels of coal dust — the main cause of black lung — in underground mines. When Lauriski arrived at MSHA, he vowed to "make a culture change" in the agency, shifting its emphasis from enforcement to education, training, and consulting. (Mine inspectors, for example, became known as "health and safety compliance officers.") As if that weren't enough, Lauriski's boss, Secretary of Labor Elaine Chao, was the wife of Kentucky Republican senator Mitch McConnell, one of Big Coal's staunchest supporters. And no wonder: according to Common Cause, from 1997 to 2000, the coal industry gave $584,000 to the National Republican Senatorial Committee, the soft-money machine that McConnell then chaired. Despite his insistence that there was no conflict of interest, McConnell left his fingerprints all over his wife's agency. Chao even hired several of McConnell's Senate staffers to work in top positions at the Department of Labor, including Chao's chief of staff, Steven Law, who had been one of McConnell's top aides.

Coal operators were not bashful about exploiting these political connections. In 2002, notoriously outspoken Ohio mine owner Bob Murray threatened to have some MSHA inspectors fired because they had cited his mines for failing to comply with regulations. "Mitch McConnell calls me one of the five finest men in America," Murray told the inspectors, according to local press accounts. "And the last time I checked, he was sleeping with your boss."

It wasn't long before tragedy struck again. On September 23, 2001, as many Americans watched the World Trade Center memorial service at Yankee Stadium, a crew of thirty-two miners descended into the deepest coal mine in North America, near Brookwood, Alabama. The Jim Walter Resources No. 5 mine was twice as deep as the World Trade Center had been high, and it was one of the most hazardous mines in the country, with a long history of safety violations, many of which related to the buildup of dangerous methane gas. At about 5:15 P.M., a fifty-six-year-old miner named Gaston Adams Jr. and three others were working to shore up an unstable roof. When part of the roof collapsed, a falling rock hit a sofa-size battery charger, sparking a methane explosion. When the dust cleared, Adams was pinned beneath the debris. He gave his headlamp to his buddies and told them to get out. But when news spread that Adams was trapped, twelve of his fellow miners decided that instead of evacuating, they would try to save him. It was a heroic, but fatal, decision. Forty-five minutes later, a second, more powerful blast killed them all and set off a fire that raged for three days, until the company flooded the shaft. It was the worst coal mining disaster in nearly twenty years, and yet because it happened just twelve days after 9/11, most Americans never heard of it.

After investigating the explosion in the mine, which was nonunion, the United Mine Workers accused MSHA of treating "serious violations such as . . . disruptions in the mine's ventilation system" as "minor infractions." The union claimed that MSHA did not respond to "requests by the miners for increased inspections when serious hazards existed," that the agency provided the mine owner with advance notice of inspection locations, and that an "MSHA supervisor divert[ed] an inspector away from an area of the mine that had known ventilation problems just prior to the explosion." At the time of the disaster, the mine had thirty-one outstanding violations, and federal inspectors had not bothered to determine whether they had been corrected. Just six days before the explosion,

MSHA inspectors wrote that methane concentrations posed an "imminent danger." Why was nothing done? One surviving miner, Robert Tarvin, pointed to the cozy relationship between MSHA and Jim Walter Resources: one of the MSHA officials to whom mine inspectors reported had recently been a manager at Walter Resources. "Going out for dinner [together] and playing golf and this junk," Tarvin said with a scowl.

"If you know the history of this industry," observed Joe Main, the health and safety director for the United Mine Workers, "you can see why miners would be concerned when coal company bosses take over the administration of mine safety and health laws."

After the explosion in the Sago mine in 2006, questions were again raised about MSHA's competency and its willingness to enforce mining laws. "Where is MSHA?" West Virginia Senator Robert Byrd asked during a fiery Senate floor speech not long after the bodies of the twelve men were pulled from the mine. "What is that agency waiting for?" Byrd complained that the Sago mine had received 276 MSHA citations over the last two years but was still allowed to operate. "Could an automobile driver or a truck driver rack up 276 speeding tickets and still have a license? What if someone had 276 mistakes on a tax return?" Byrd asked. "But here was a coal company with 276 violations and still operating."

In the aftermath of the tragedy, Congressman George Miller, the ranking Democrat on the committee that oversees MSHA, called on the Bush administration to dramatically increase fines against mining companies that repeatedly violate federal safety rules. In a letter to Labor Secretary Chao, Miller pointed out that MSHA fined the owners of the Sago mine just $24,374 for the citations issued in 2005, or an average of only $156 per violation. Many of these citations, Miller noted, "were the result of repeated violations of the same rules and regulations, over and over." For example, one citation in December 2005 for accumulation of combustible material received a $60 fine. It was the twenty-first citation that year for the same type of violation. "In such a profitable industry, there is no reason to tolerate repeat violations by mine operators," Miller wrote in his letter to Chao. "When a speeding ticket in West Virginia costs more than the twenty-first citation for accumulation of combustible materials, there is something horribly wrong with mine safety enforcement."

If a doghole like Quecreek is the lowest rung on the coal miner's ladder, a strip mine like Cordero Rojo in Wyoming is the highest. In Wyoming, the saying goes, they don't really mine coal; they just move dirt. And during my time in the Powder River basin, no one I met moved more dirt than Joel Hjorth. At Cordero Rojo, Hjorth operated the $60 million Marion 8200 dragline, one of the largest pieces of heavy equipment in the world. It is essentially a huge bucket at the end of a long boom that swings back and forth, picking up dirt from one side and dumping it on the other. It can move seven thousand yards of dirt in a ten-hour shift.

Hjorth looks more like a cowboy than a coal miner: big silver belt buckle, tight Wranglers, and a bit of a swagger. Hjorth's job is to sit in a big air-conditioned cab on the dragline all day and, using hand and foot controls, command the bucket to move dirt from one side of the mining pit to the other. This takes a lot of dexterity and judgment, but very little muscle. When I met him, Hjorth had been operating the dragline for eight years — it was all second nature to him. While he worked, he often blasted Led Zeppelin in the cab. Hjorth sometimes left after a day's work with hardly a speck of coal dust on him.

It's a simple truth that if you're going to work in a coal mine, Wyoming is the place to do it. With a little overtime, an operator like Hjorth can make $60,000 a year. With a lot of overtime (and, at the very least, a lot of caffeine), he can make $90,000. Haul truck drivers and bulldozer operators are paid less, but they generally are more highly paid than people in comparable jobs in the East. Working in a coal mine in Wyoming is also much safer than it is anywhere else in the country. In part this is because there are no underground mines in the Powder River basin. In addition, the corporate owners of the Wyoming mines often put a higher priority on safety than some of the smaller, short-sighted operators back east.

In Wyoming, miners like Hjorth often joke that their jobs could be done by robots. But it's not entirely a joke. In recent years, the explosion of high-tech equipment in big strip mines has been mind-boggling. At a big mine like Cordero Rojo, haul trucks and bull-dozers are tracked via GPS; each machine's whereabouts can be displayed on a computer screen anywhere in the mining office. A dragline operator's every move is recorded on a computer, and at the end of each week, the operator is presented with a chart that shows precisely how much dirt he or she moved, and how he or she ranks compared to other operators at the site. All this may help to make the mines incrementally more efficient, but it makes many workers feel like drones. I have met many miners in Wyoming who say that they are treated fairly and respectfully in their jobs, but I have never met one whose eyes lit up like Randy Fogle's when he talked about his work underground. Fogle was making half as much money as the miners in Wyoming, in conditions that were five times more dangerous, but he was fifty times more alive in his job and, I'd estimate, a hundred times happier.

Like coal companies everywhere, Wyoming companies are about to confront a demographic nightmare. The average age of a coal miner in the United States is fifty years old. Unless the mines start attracting younger people, the coal industry is going to die of old age. In Wyoming, the mines have been staffed largely by workers who started in the 1970s, when the mines opened up, creating thousands of new jobs. Now those workers are retiring, and the industry is scrambling to replace them. In 2005, some coal companies began offering signing bonuses and other perks to attract young workers. Others have sought changes in state laws to allow the hiring of migrant Hispanic workers in coal mines (for safety reasons, current regulations in most states require coal miners to read and speak English).

Even in coal country, mining jobs are a tough sell. Kids don't grow up with Tonka trucks anymore; they grow up, for better or worse, with Halo 2. In addition, coal mining is a very conservative, very white, and very uncreative business, even compared to other extraction industries, such as oil and gas. And mining companies are hardly known for pampering their employees. In recent years, they have turned their backs on thousands of retired miners, fighting for regulations that make it increasingly difficult for workers to receive black lung–related medical benefits. No wonder mining programs at universities and colleges are withering. Nationwide, in 2004 only about 100 students graduated with mining engineering degrees; 65,000 graduated with law degrees. Christopher Bise, the head of the undergraduate mining engineering program at Penn State, one of the most prestigious programs in the nation, believes that the coal industry has only itself to blame. "They've taken their workers for granted for a hundred years," Bise says. "Now they wonder why no one wants to go to work for them?"

The breakthrough at the Quecreek mine happened at about 9 P.M. on Wednesday, July 24, 2002. At the time, Mayhugh was running the shuttle car, hauling coal from the coal face to the conveyor belt. Sixty billion tons of water flooded in, creating a raging river that quickly began to fill up the mine. Although Mayhugh was arguably in the best physical shape of all the miners, he was also the most emotional. During their seventy-seven-hour ordeal, Mayhugh tried to stay sane by cracking jokes and building walls to keep the water back. But when the water kept rising and it was time to write farewell notes to their families, Mayhugh was among the first to break down. It was the more experienced miners, especially Fogle, who kept them on an even keel.

In retrospect, there had been signs that they were headed for trouble. The biggest worry had been water leaking in through the roof of the mine. This in itself was not unusual, nor was it necessarily a sign that they were approaching the flooded Saxman mine. But that thought must have crossed Fogle's mind. Water pumps were brought in to dry out the mine, but they couldn't keep up. Several miners felt that something wasn't right. John Unger, one of the roof bolters on the same crew as Mayhugh and Fogle, later remembered saying aloud, "This place is going to turn to shit shortly."

It was Fogle's job to communicate any worries or concerns that the miners might have to the mine superintendent. In later testimony, Joe Gallo, the vice president of operations at PBS, denied that Fogle had said anything to anyone in management about deteriorating conditions in the mine. In public statements, Fogle also has denied that he ever expressed concern to anyone at PBS, including Dave Rebuck, the subcontractor who was in charge of mining operations at Quecreek. Fogle says that he did advocate getting out of there, but only because the water was slowing up production so much that it hardly seemed worthwhile to continue working. When they were trapped underground, however, one miner says that Fogle told him that he had had conversations with Rebuck about the conditions in the mine and that Rebuck, according to Fogle, was "scared."

For legal reasons, Fogle was not willing to discuss exactly what he did or didn't say to Rebuck about the conditions in the mine. But the larger, and in some ways more significant, point of all this is that workers in a mine like Quecreek learn to keep their mouths shut. With no union to protect them, and in an area where hundreds of out-of-work men are waiting in line for jobs, a miner who complains about dangerous conditions will soon find himself stocking yo-yos at the Wal-Mart. For a person like Fogle, taking a stand against management essentially meant putting his career on the line. In fact, he was already in trouble with them. A year or so earlier, he had had run-ins with a mine boss at PBS that had gotten him demoted from superintendent to shuttle car operator and had resulted in a 50 percent pay cut. Was his sense of danger strong enough for him to warrant putting his job on the line again? He had to know that if he did complain, he might never work in a coal mine in Somerset County again. Fogle is a man of tremendous integrity and would never have sent his crew into a mine that he knew was dangerous — but what if he was only 80 percent sure that it was safe?

As for the mine inspectors, many of them were regarded by the miners as a joke. Often they were friends with the mine owners. "We all know when they're coming," Tom Foy told me. "We'd pretty the place up for a few hours when they came through, then we'd get back to work."

The first night the miners were trapped underground, when it was still unclear to those outside whether they were dead or alive, I spent several hours at the rescue site talking with two experienced mine inspectors from the Pennsylvania Bureau of Deep Mine Safety about mining accidents in the past. One of the inspectors told me that when he heard about the breakthrough, the first thing he thought of was the Knox mine disaster — a famous coal mining tragedy in the anthracite fields of northeastern Pennsylvania in 1959.

In fact, the parallels between the two disasters are eerie. Both played out against a backdrop of economic decline. By the 1950s, as the anthracite seams in northeastern Pennsylvania became depleted, big corporations were no longer willing to invest capital in the mines and began selling out to smaller companies. These companies hired independent contractors to work the mines, which, among other things, released the companies from liability. They busted unions, paid off regulators, and stole coal from old mines. Eventually, the region attracted the likes of Jimmy Hoffa, the head of the Teamsters union, who was a secret investor in one of the last companies to try to squeeze a profit out of the mines. (Hoffa may have had a soft spot for the industry because his father was a coal miner who died young from black lung.)

On January 22, 1959, miners taking the last bit of coal out of a seam beneath the Susquehanna River heard a loud crack. "Water poured down like Niagara Falls," one miner remembered. For several days, the river flooded into the mine, creating a giant swirling whirlpool on the surface of the river. Panicked mine engineers dumped thousands of tons of rock and dirt into the hole in the riverbed but failed to stanch the flow. Finally, they built railroad tracks to the edge of the river and pushed in hundreds of mine carts, as well as a number of full-size railroad cars, which clogged the hole in the river bottom. Twelve miners died, their bodies forever entombed beneath the river. The disaster essentially killed deep mining in the area.

What would be the legacy of Quecreek? "That all depends," the mine safety expert speculated, "on how many bodies they pull out of the ground. I don't want to make any predictions, but if this goes badly" — he glanced toward the frantic workers setting up the drill rig that would ultimately cut the hole through which the miners were rescued — "it could end coal mining in this part of the country."

His fears, of course, were unfounded. Nevertheless, in the immediate aftermath of the rescue, most of the trapped miners, especially

Mayhugh, were under the illusion that someone would be held responsible for what had happened to them. No one could construe this, to borrow the favorite excuse of Big Coal, as "an act of God." They had run into an old mine that everyone knew was there and that everyone knew was flooded with water. How could that have happened? "Clearly, someone screwed up," Tom Foy told me not long after the rescue. "We are gonna find out who and why." Pennsylvania governor Mark Schweiker, who spent a lot of time in front of the TV cameras at the rescue site and had come to visit the miners in the hospital, had guaranteed them as much. He appointed a blue-ribbon panel and promised, as smart politicians in similar situations always do, to get some answers. In addition, the state DEP opened an investigation, as did MSHA.

It didn't take Mayhugh and the others long to figure out something was up. Within a few weeks, Governor Schweiker announced that his blue-ribbon panel was not going to get any answers after all. Instead, their job would be to make recommendations to prevent this sort of "accident" from happening in the future. You could smell the whitewash a mile away. The investigation by the DEP was equally toothless. The state investigators could hardly be bothered to go through the motions of an investigation. In some cases, the investigators' interviews with the miners lasted only twenty minutes, with much of that time taken up by introductions and legalese, and the rest of it consisting of softball questions about who did what in the mine. In addition, a lawyer for the coal company was allowed to be present during the miners' questioning — a gesture of intimidation that virtually guaranteed that nothing critical would be said of the coal company or the way it ran the mine. Of course, the miners themselves bore part of the blame for the lack of candor in these interviews. When offered a chance to speak their minds, they rarely did, in part because they were afraid of legal complications, in part because they didn't want to betray their brethren, and in part because, as beaten-down coal miners in a beaten-down industry, they were accustomed to shutting up before people of authority.

In any event, when the state investigation was released, it essentially blamed the whole thing on bad maps. The MSHA report was slightly more critical, citing both PBS Coals and the engineering firm that had certified the Quecreek mine map for "moderate negligence" for not checking production records and other documents from the old mine that would have indicated that the maps were inaccurate. Total fines: $14,100. It wasn't even a slap on the wrist.

Among the many unanswered questions was this: did PBS Coals indeed have a better map of the old Saxman mine but refuse to disclose it? If it did, that was a potentially criminal act. In a report to Kathleen McGinty, head of the DEP, an investigator in the inspector general's office noted that Tom McKnight, a state mine inspector, had been told by Joe Gallo, a vice president at PBS Coals, that the company had a more recent map of the old mine than the one they had filed with the permit. McKnight asked for it repeatedly, at one point even "jumping on" Gallo to produce it, but Gallo never did. After the accident, Gallo denied that he had ever had possession of the map. A grand jury investigation determined that it was essentially McKnight's word against Gallo's, and that there was insufficient evidence of criminally reckless misconduct on the part of PBS Coals, Mincorp, or its contractors. In 2005, when I contacted McKnight to confirm these events, he would only say, "Someday, the truth will come out about all this."

Strangely enough, two more recent maps of the Saxman mine turned up after the rescue: the first appeared a few weeks after the breakthrough, at a coal museum in Somerset County; the second was found in late 2004 in a box at Consol Energy, which had owned the Saxman mine in the 1970s. Both maps had more accurate renderings of the boundaries of the old mine. Had they been used by PBS, they might well have prevented the breakthrough.

What motive might PBS Coals have had to conceal a more up-to-date map? Simple economics: the more inaccurate the map, the greater flexibility a company has to mine more coal. A cowboy coal company might gamble that an old map would allow it to squeeze out an extra few thousand tons of coal before it hit the old mine. And if the company did hit the mine? Well, coal companies have been gambling with workers' lives for a hundred years. It was nothing new.

For Mayhugh, the results of the investigations were a bitter disappointment. "It really opened my eyes to how the system works,"

Mayhugh told me. "It's not about protecting the miners, that's for sure." The very people and institutions he had previously trusted, from the mine owners to the state inspectors, were the ones least willing to stand up for him.

To make matters worse, when Mayhugh and other miners publicly raised questions about the thoroughness of the investigations, they were treated like turncoats by fellow coal miners and other people in the community: *Just be grateful and shut up* seemed to be the message. Undaunted, they hired a lawyer to go after the coal company on civil charges of negligence and reckless endangerment. Eight of the nine miners eventually joined the suit, which, as of this writing, had still not gone to court. "I'm not taking them to court because I want money," Mayhugh told me, sitting in the living room of his house about a year after the rescue. "I'm doing it because I want the truth."

In the end, the breakthrough at Quecreek cost PBS Coals almost nothing. Pennsylvania taxpayers ended up covering the $1.5 million tab for the rescue operation and reimbursing PBS more than half a million dollars for losses and fees related to the rescue. After closing for a few months to pump out the water, the Quecreek mine opened again. The year 2004 turned out to be a banner year for the company — the best since the heyday of the 1970s. But safety remained dubious. In early 2005, two workers at Quecreek were hospitalized, one with critical burns, when a transformer blew up in their faces. A month later, another miner was hospitalized after suffering an electrical shock. Meanwhile, MSHA chief Dave Lauriski resigned his post shortly after the 2004 election under a cloud of scandal. During his tenure, the Department of Labor's inspector general concluded, MSHA had wrongfully awarded no-bid, single-source contracts totaling more than $500,000 to two companies with ties to Lauriski and one of his lieutenants. Nobody was surprised when, shortly after his departure from MSHA, Lauriski landed a new job as a highly paid coal industry consultant.

Randy Fogle was the only one of the Quecreek miners who went back to work underground. He returned to the same mine, in fact, wearing the same hardhat he had worn the night the water rushed in. There's something both inevitable and elegiac about Fogle's return to the darkness. He understood very well that men like himself who love the challenge of mining coal are fading fast from American life. "In the Fogle family, I'm the end of the line," he told me one afternoon as we stood out in his backyard, admiring his wife's tomato plants. "My boys are going to have to find something else to do with their lives."

As for Mayhugh, it took him a couple of years to figure out what he wanted to do next, but eventually he made a clean break with his past. He now works as a maintenance engineer at a Pennsylvania wind farm.

Chapter 4

The Carbon Express

ONE VIRTUE OF COAL as a source of power is the fact that it can be stockpiled in silos or, if necessary, in big piles on the ground. Over time, exposure to the elements will degrade coal, causing it to lose its volatility. Left alone for too long, it can also begin to smolder and burn — or even explode, if the conditions are right. But if you handle it right, coal can be stored for a few months, maybe more. Most coal plants in America keep a thirty- to forty-day supply on hand and employ half a dozen people to tend the coal pile — moving it around with bulldozers, watching for smoldering or smoke, keeping it well mixed and fresh. In this sense, coal has a big advantage over its main rival, natural gas, which is delivered to power plants via big pipelines on a just-in-time basis. Any interruption — a break in the line, a malfunctioning valve — is catastrophic for the smooth operation of the power plant. In contrast, when you burn coal, you know exactly where it is and how much you have. "A big pile of coal," one Florida power plant manager told me, "is like money in the bank."

But in other ways, the fact that coal is a rock is a major drawback. Unlike wind and solar power, where nature (or God, depending on your point of view) delivers the fuel source, one of the problems with coal has always been that you have to haul it from the places where it is mined to the places where it is burned. One of the big reasons that oil displaced coal as a fuel source in the first half of the twentieth century is simply that oil is liquid — it can be piped, pumped, processed, and stored with relative ease. Moving coal is a dirty, difficult, expensive endeavor. The problem of how to transport it efficiently is nearly as old as our romance with coal.

Canals were the first solution. In the 1700s, horses pulled heavy wagons out of British coal mines on muddy trails to the water, where it was loaded onto boats and shipped around the country. Soon wooden rails were laid down on the bare earth, greatly improving haulage speed to the canals. In 1825, horses were dispensed with altogether when the man known as the father of the railroad, George Stephenson, opened the first modern railway between the British coal town of Darlington and the river town of Stockton. The locomotive was a coal-fired steam engine; in effect, coal hauled itself. The partnership of railroads and coal created a kind of perpetual motion machine: better transportation meant cheaper, wider distribution of coal, which fed the growth of steel mills and steam power, which in turn further increased the demand for coal. The partnership lasted for more than one hundred years, until railroads replaced coal with diesel fuel, which was more efficient and easier to handle. But coal remains as dependent on the rails as ever.

Nowhere is that clearer today than in the West. The eighty-foot seams of coal in the Powder River basin would be worthless without reliable, efficient transportation. Over the past thirty years, the two major railroads that haul coal out of the basin — the Burlington Northern Santa Fe (BNSF) and the Union Pacific (UP) — have spent hundreds of millions of dollars to build one of the most technologically sophisticated rail networks in the world. Because of the speed and efficiency of the railroads, Wyoming coal is now burned as far east as Massachusetts, displacing dirtier Appalachian coal and creating what amounts to a nationwide conveyor belt that is largely responsible for keeping coal cheap and plentiful. The amount of train traffic that flows in and out of the Powder River basin is staggering: the BNSF alone hauled about 250 million tons of coal out of the basin in 2005 — roughly fifty mile-long trains each day. The UP, initially shut out of the Wyoming coalfields, finally gained access after buying out a smaller railroad in the early 1980s and now hauls even more coal than the BNSF. Today, depending on your point of view, the UP and BNSF are either bitter competitors or quietly sharing a very lucrative coal-hauling monopoly.

A good percentage of the coal traffic flowing out of the Powder River basin rolls through Edgemont, South Dakota, a former uranium boomtown on the edge of the Black Hills. Edgemont is 110 miles southeast of Gillette, Wyoming, and the first stop for BNSF trains heading out of the Powder River basin. From Edgemont, the trains roll on to Alliance, Nebraska, which is 121 miles to the southeast, over a mean grade known as Crawford Hill. These 231 miles of track between Gillette and Alliance carry more tons of coal per mile than any other stretch of rails in the United States. Some thirty loaded coal trains, each more than a mile long and each carrying more than ten thousand tons of coal, pass along this route every day. It is one of Big Coal's most vital — and most fragile — delivery routes.

One recent winter morning, I stopped in at the Edgemont station, hoping to hitch a ride on an eastbound coal train. (I had contacted BNSF's main office in Fort Worth, Texas, a few days earlier and received written permission to ride along.) I bumped into Moon Taylor and Dave Anderson, a BNSF train crew waiting for Engine 9722 from Gillette. Taylor, the conductor, told me that 9722 — the lead of three locomotives on the train — was pulling 125 cars, each loaded with about one hundred tons of coal from Kennecott Energy's Jacobs Ranch mine.

Taylor looks more like a Little League coach than a railroader. He is thirty-five, lean and strong, with blue eyes and a mustache. He wore a faded blue baseball cap with a crescent moon on the front and the word "Moon" on the back. He's from an old railroading family; his father and grandfather were both brakemen. While we waited, Taylor double-checked the train's manifest on his Sony handheld: total weight, 17,891 tons (about 12,500 tons of coal); total length, 6,857 feet; scheduled arrival time, 8:12 A.M.

While Taylor fussed with his handheld, Anderson, the engineer, wolfed down a breakfast burrito and a cup of coffee. Anderson is tall and a little ungainly, with black hair and eyes and an easy smile. He is a full-blooded Sioux, given up for adoption by his birth mother and raised by a Lutheran minister. As I would later learn, Anderson is well-known among railroaders in the area, not only because he's the only Sioux hauling coal in Wyoming, but also because his older brother, Tim, who was also an engineer, was killed in a controversial train wreck in 1994.

Anderson had just finished his burrito when a distant whistle sounded.

"Is that it?" he asked Taylor.

Taylor looked at his watch. "Must be. And only ten minutes late."

We pulled on our gloves, donned our hardhats and safety glasses, and stepped out into the cold to greet 9722.

Railroads love coal, mostly because it is high-volume, low-hassle business. There are no stops to unload freight cars, no hazardous chemicals, no passengers, just an endless back-and-forth between the mines and the power plants. In the East, railroads face competition from river barges and, to a lesser degree, trucks. But in the West, trains rule. More than 20 percent of BNSF's annual revenues are from coal, with the vast majority of that coming from the Powder River basin. Here's why: in 2004, a ton of Wyoming coal delivered to a power plant in Georgia sold for about forty dollars. Of that, as much as 80 percent went to the railroads. They are, as one coal executive put it to me, "the monster in the middle."

The success of the railroads, which allowed power plants in the Midwest and Southeast to switch from expensive, high-sulfur coal mined in central Appalachia and Illinois to cheaper, low-sulfur Wyoming coal, has led to lower emissions of sulfur dioxide (a pollutant that causes a variety of health effects, as well as acid rain) and lower electricity bills for millions of Americans. By any measure, that is a good thing.

But the spread of Wyoming coal has had other, less obvious consequences. Because Wyoming coal has a lower heat value than most eastern coal, power plants have to burn more of it, causing carbon dioxide emissions to rise when plants switch to Wyoming coal. Also, the spread of Wyoming coal has put intense price pressure on Appalachian coal companies, adding further incentive to close union mines and pursue the lowest-cost mining techniques, such as mountaintop removal. Most important, the success of Wyoming coal has meant that BNSF and UP are the 800-pound gorillas of the coal industry, with a virtual lock on the movement and price of nearly half the coal burned in America. This fact only underscores one of the most disturbing — or, depending on your point of view, rewarding — aspects of the coal business: the control of the industry by ever fewer, ever bigger corporations.

The power of the railroads also has been boosted in recent years by their close ties with the Bush administration. Matt Rose, the forty-six-year-old CEO of BNSF, was a Bush Pioneer fundraiser (a supporter who brought in at least $100,000 in campaign contributions) in 2000 and 2004. Marc Racicot, the chairman of the Republican National Committee and head of President Bush's reelection campaign in 2004, has been a BNSF board member since 1996. And John Snow, President Bush's Treasury secretary, is the former head of CSX Corporation, the Florida-based railroad that hauls millions of tons of Appalachian coal. Not surprisingly, the railroads are all big supporters of coal and have used their influence to help kill any legislation that might cut consumption. The railroads have been especially helpful in derailing legislation that might tax or regulate carbon dioxide emissions. BNSF was a supporting member of the Global Climate Coalition, an alliance of corporations whose mission was to undercut, obfuscate, and confuse the science of climate change. (Formed in the early 1990s, the coalition at one time also included ExxonMobil and a number of coal and electric power companies.) In addition, until recently, BNSF was a key contributor to the Greening Earth Society, a group largely funded by the coal industry to promote the idea that global warming is good for the planet because it allows crops and trees to grow faster. Like many Big Coal executives, BNSF's Rose has been very good at using economic threats to argue against any proposal (such as the Kyoto Protocol) that would place limits on carbon dioxide emissions. "I think the future looks bright, with one exception," Rose told a gathering of BNSF employees in late 2003. "If an environmental bill is passed [that caps carbon dioxide emissions], employment would be cut by half or more, because coal would be seriously hurt or even eliminated."

Such rhetoric is not unusual from coal executives such as Massey Energy's Don Blankenship, but it is downright hypocritical coming from the head of a railroad. BNSF's own publicity materials boast about how rail service is superior to trucking because air pollution and carbon dioxide emissions are much lower per ton of freight moved. This is entirely accurate, and a good reason to choose railroads over trucks, but to make this argument while simultaneously kneecapping climate change legislation is Orwellian. The truth is, environmental laws have been very good to BNSF. After all, it was the 1970 amendments to the Clean Air Act that set limits on the sulfur content of coal and thus created the market for Wyoming coal from which the railroads are now profiting so handsomely.

For all their political power and influence, however, the railroads remain the most visible symbol of the inefficient, creaking, antiquated infrastructure on which Big Coal depends. Already railroad congestion is the biggest bottleneck to the expansion of coal-fired power plants. Technologically sophisticated as the railroads may be, there is simply too much coal moving in too many directions for them to handle. Delays are increasing, as are train derailments, breakdowns, human error, and weather problems. Arch Coal, which owns huge mines in Wyoming and Appalachia, estimates that railroad congestion cost the company more than $8 million in lost revenues in the first half of 2004. The expansion of rail lines, or perhaps even an entire new rail route into the Powder River basin, is the subject of much discussion but little action. And no wonder: new track costs $3 million a mile. The whole system is so baroque that it makes you wonder: is this America's energy future? We can build iPods, send robots to Mars, and unravel the human genome, but when it comes to generating electricity, we rely on a system that requires hauling millions of tons of black rocks from one side of the country to the other. Even the railroads understand how inefficient coal is; they switched to diesel fifty years ago.

Nobody in the railroad industry seems terribly worried about any of this. The railroads were the great romantic leads (both hero and villain) of the industrial age, and their self-assuredness continues today. They believe that, ultimately, Big Coal needs them more than they need it. In fact, if you judge from the half-hidden fear and frustration in the voices of coal industry executives when they talk about the railroads, you might think that the future of coal lies entirely in their hands. "If there's one thing you don't want to do in this business," Bret Clayton, CEO of Kennecott Energy, told me, "it's piss off the railroads."

It was snowing lightly when Engine 9722 pulled into the station, its big diesel engines rumbling as it glided to a stop. A Powder River basin coal train is one of the most magnificent manifestations of sheer horsepower in the world — a mile-long parade of silver cars pulled by two big locomotives in front and pushed by one in back. The train is so long that the front locomotives are often going down one side of a hill while the engines in the rear are still pushing it up the other. I heard one conductor joke how coal trains are getting so long that they sometimes occupy three states and two time zones.

Taylor and Anderson crossed the tracks and climbed into 9722, a diesel-powered SD70MAC locomotive. Anderson tossed his duffle bag on the floor behind the engineer's seat on the right side. He pulled out his reading material — a pile of sports magazines and a Dell computer catalog — then he took out a bottle of Formula 409, sprayed it all around the controls, and wiped them down with paper towels. "You never know who was in here before you," he explained.

Taylor settled in on the conductor's side and immediately read the latest track orders, a daily computer printout that tells conductors about track conditions — where to expect delays due to track work, which sections of the route have reduced speed limits.

"We've still got the slow order at milepost 449, but otherwise we're good," Taylor told Anderson.

"Well, then, let's get rolling," Anderson said. He dropped his wraparound sunglasses over his eyes and pushed the throttle lever up to the first notch. The big engine behind us began to rumble and shake. Slowly, almost imperceptibly, we began to roll down the tracks.

Coal moves out of the Powder River basin along three routes. The northern line, which is owned by BNSF, heads northwest out of the basin into Montana, where trains can turn west toward Oregon and Washington or east toward Michigan, Wisconsin, and Minnesota. The southern route, which is shared by BNSF and UP, crosses the old Oregon Trail near Guernsey, Wyoming, then continues southwest and connects with the old UP transcontinental line in Nebraska, where it heads into Omaha and points east.

The third and most historic route is the one we were taking out of Edgemont. The line heads east out of Gillette, crosses the sagebrush prairie to the edge of the Black Hills, and then begins a long, slow climb through limestone buttes to the high plains at Alliance, Nebraska, where BNSF has a big rail yard. Besides being the busiest route into and out of the basin, it is also one of the most dramatic. The climb over Crawford Hill near the Pine Ridge Indian Reservation attracts rail fans from around the world. The route was laid out in the 1880s by Edward Gillette, a surveyor for the Chicago, Burlington & Quincy (CB&Q), who plotted the line in the shadow of the retreating Sioux and Cheyenne. (The CB&Q recognized his work by naming the town of Gillette, Wyoming, after him.) Today the tracks are still littered with remnants of the early days of American railroading: abandoned hotels, stockyards, and loading ramps, and the faint outlines of lakes created to hold water for steam engines.

Riding a coal train in the West is nothing like riding Amtrak up the Hudson River in the East. For one thing, there is no clackity-clack of the rails, because the heavily trafficked coal-hauling routes all use ribbon rails. Unlike traditional rails, which are bolted together to give the steel room to expand and contract, ribbon rails are welded together into a single seamless line of steel. Ribbon rails occasionally snap in the winter, when the cold causes the steel to contract, or turn wiggly in the summer, when the heat causes the steel to expand. But they can handle far more weight than traditional rails, cause less wear and tear on the trains, and give an exceedingly smooth ride.

Nevertheless, getting 12,500 tons of coal from point A to point B is never a simple matter. The day before, the coal train I was riding developed problems with the air brakes, causing them to grab unexpectedly. We were marooned near Newcastle, Wyoming, for six hours. Another train I rode was held up for four hours because of track congestion. In fact, of the half-dozen coal trains I'd ridden in the past few days, every one of them had been plagued by long delays or mechanical problems. You didn't need a team of consultants to see that the system was stretched to its maximum capacity. And the problems weren't all mechanical. One afternoon, as I walked back from lunch with Will Cunningham, a coal marketing executive for BNSF, we came across several workers staring at a smashed window in the side door of BNSF's Gillette station. Obviously, someone had just slammed his or her fist through the glass. "What happened?" Cunningham asked. "Same old shit," one of the workers replied. "Trying to do too much with too little. Somebody got pissed off about it and vented their frustration."

In Engine 9722, our problems began almost as soon as we left the station. Just outside Edgemont, as the route cuts across the southwesternmost corner of the Black Hills, the tracks begin a gentle rise, hardly even a hill. As we headed up it, the rumble of the engine grew louder and lower, and the cab began to shake.

Anderson said, "She's not pulling."

Taylor glanced at the speedometer: it was falling rapidly — 20 miles per hour, 19, 18, 17. Anderson moved the throttle to notch 8, full throttle, but the train kept slowing. The steel wheels spun against the rails, squealing. Anderson checked the gauge that shows the distribution of power among the three locomotives. "Number two engine feels like a lame duck," Anderson said. The three locomotives are wired together via computer to distribute the power equally (the railroad equivalent of four-wheel drive). If number two wasn't pulling, the lack of horsepower would put additional strain on the knuckles, or connectors, between the cars, causing them to snap like cheap plastic toys.

"We're gonna bust, I can feel it," Taylor said, shaking his head. If the knuckles snap, it's the conductor's responsibility to jump out and replace them — not an easy job under the best of circumstances, and a downright miserable job on a cold day like this.

"We might make it," Anderson said hopefully.

The train slowed to 7 miles per hour. The crest of the hill was just ahead, but it was not at all clear that the train would have the strength to make it over the top. We slowed to walking speed. We could feel every molecule of steel in the train straining to its breaking point.

Taylor braced himself. "Hold on," he said to me. "When this thing breaks, it's gonna lurch forward. It'll throw you through the windshield if you're not ready for it."

I held on. But there was no need. Finally, wheezing and shaking like the Little Engine That Could, 9722 crested the hill. Anderson called the BNSF dispatch center in Fort Worth to report the problem, while Taylor slouched back in his seat, relieved that, for the moment anyway, he wouldn't be out there toting iron. As we headed down the other side, gravity took over, and soon we were gliding at 50 miles per hour through the high grasslands of South Dakota.

The railroads' near-monopoly over coal hauling is important in three ways. First, it allows railroads to more or less determine the price and availability of coal. Second, because the railroads are bullies, only the biggest customers have the leverage to negotiate with them, which means that smaller customers — both mines and power plants — tend to get forced out of business, concentrating power in fewer and fewer hands. Finally, major congestion or a derailment on key railroad lines can have profound consequences for both mines and power plants. In 2005, after coal shipments were delayed because heavy rains caused back-to-back derailments on a main line in the Powder River basin, coal delivery was down about 20 percent for months, and several power plants in the Midwest nearly ran out of coal. Big Coal executives love to tout coal plants as being invulnerable to terrorist attacks (at least not in the way that nuclear or even gas plants are), but these same executives will tell you privately that the destruction of a single railroad bridge over the Mississippi River could have a disastrous impact on power plants in the East.

Although coal miners and power plant operators admit their dependence on the railroads, they are not happy about it. In fact, if you spend any time at all around coal-hauling railroads, you will quickly become aware of a paradox: Among outsiders, railroads are almost universally loved for their power, their efficiency, and their ability to conjure up memories of the model choo-choo trains that ran around our childhood Christmas trees. Among insiders, railroads are viewed with a mix of fear and awe, reviled for their arrogance and political influence and considered by many to be nothing less than a network of heavy-metal bandits.

To understand this paradox better, before I left the Powder River basin, I spent a few hours in Cordero Rojo's loading silo, where the coal is dumped into railroad cars. Coal silos are the noisiest, dustiest, dirtiest, and most controversial places at any coal mine — even though what happens there is pretty simple. As an operator in a glass-fronted booth operates the controls, thousands of tons of coal rumble out of the silo and into the coal cars waiting below. It comes down like thunder, dust billowing up for a second and obscuring the daylight. The train below never stops, just rolls through at about 2 miles per hour, until all the cars — usually 130 or so — are loaded. Then the next train, most often already waiting on the loop line, pulls in.

But exactly how much coal is loaded onto each train? The cars are weighed as they come in and again as they go out, so the railroad knows — or is supposed to know — precisely how many tons of coal are in each car. This is important, because the power companies pay for the coal based on what it weighs at the mine. When you're buying millions of tons of coal a year, small errors compound quickly. And some power companies suspect that the scales are not as accurate as they might be. The Southern Company, an Atlanta-based utility that is one of the biggest coal consumers in America, is so wary of the railroad scales in Wyoming that it keeps a full-time employee, a coal industry veteran named Greg Henshaw, stationed there to monitor them.

Henshaw knows all the games the railroads play with the mines and the power companies. He compares the relationship between the railroads and the power companies to the relationship between the bull and the cowboy at a rodeo. "The only control the cowboy has over the bull is before you open the gate," Henshaw said. "After that, you just hold on for dear life. For about eight seconds, you and the bull are one." The main problem, Henshaw says, is that the coal-hauling railroads answer to no one. That, of course, has essentially been the charge against the railroads for the past 150 years. There have been various attempts to remedy this, including regulatory oversight by several federal agencies such as the Surface Transportation Board, which regulates shipping rates, but these measures have not tamed the railroads.

The unreliability of the scales is not the power companies' only concern, Henshaw explains. A far bigger problem is the complex and seemingly arbitrary rates the railroads sometimes charge to ship coal. Power plants that happen to be served by two railroads, or, even better, a railroad and a barge company, can get competitive bids for coal delivery from two companies. Power plants that are served by only one line — like the Southern Company's Plant Scherer near Macon, Georgia — are much more vulnerable. Southern pays what the railroads charge, or the plant doesn't get the coal. One afternoon, I asked the general manager at Plant Scherer, Danny Morton, why, on that particular day, electricity at Scherer's sister plant, Plant Miller in Alabama — which was built at about the same time, uses most of the same technology, and burns more or less the same coal — was dispatching its electricity for about five dollars per megawatt-hour less than Scherer. "The railroads," Morton replied. Plant Miller, he pointed out, is a few hundred miles closer to the mines in Wyoming, but it is also served by competitive railroads. "It makes all the difference," he said.

Power companies without Southern's market clout are at an even bigger disadvantage. One example is the Rodemacher power plant in Louisiana, owned by Lafayette Utilities System, a small local power company, which converted to Powder River basin coal in the 1990s. The problem is not that the Rodemacher plant is buying coal from a mine 1,500 miles away. It is that the last 19 miles of the rail line, from Alexandria, Louisiana, down to the plant, is served by only one railroad, the UP. If the Rodemacher plant wants coal, it has to go through the UP. By controlling the last 19 miles, the UP effectively controls the entire route from Wyoming to Louisiana. And, unlike the Southern Company, Lafayette Utilities System is such a small customer that it has no leverage to negotiate a better price. Terry Huval, the director of Lafayette Utilities, testified before the U.S. House of Representatives that the rates to the Rodemacher plant were 50 percent higher than those for similar distances that are competitively bid, and that this difference is costing his utility $5 million to $6 million a year.

From the power companies' point of view, the problem with price gouging by the railroads is that the only recourse is to file a complaint with the Surface Transportation Board. These cases cost several million dollars to file and involve tens of thousands of pages of complex documents. To win a case, the complainant must prove that the rail rates it is being charged are 180 percent of market value. (Legally, the railroads are allowed to charge up to 150 percent of market value.) "These lawsuits are very difficult to win and almost never worth the time and trouble," Charles Rucker, head of fuel services for Georgia Power, a Southern Company subsidiary, told me. In early 2005, concerns about abusive pricing of western coal by the BNSF and UP were serious enough to warrant an investigation by the U.S. Department of Justice's Antitrust Division. As of this writing, the investigation was still ongoing.

Power companies that fight back against the railroads are often sorry they did. "They know how to make your life miserable," Henshaw told me. Not only can they jack up prices during the next contract negotiations, but they also can start examining the company's trains more thoroughly at inspection points. As a result, trains might be sidetracked, or "lost," for a few days. Before long, instead of taking ten days for a load of coal to travel from the coal mine to the power plant and back, it might take eleven or twelve. That's why most power companies tend to put up with the railroads' shenanigans. As Henshaw put it, "If you want your coal, the [railroads] are a necessary evil."

Outside Ardmore, South Dakota, in the middle of a stretch of empty grassland, we passed a middle-aged man in a baseball cap parked on the side of the road. He had a camera set up on a tripod, pointed at the train. Taylor spotted him first. "We got a foamer up here," he said. "Foamer" is railroad slang for a person who gets unduly excited by the sight of a moving train.

As we rolled by, Anderson blew the horn, and Taylor opened the window and waved.

I asked Taylor if he considers himself a foamer. He does, after all, come from three generations of railroaders. "No way," he said. "To me, a train is just a heavy piece of machinery — no more and no less."

On coal trains, in fact, human beings are more trouble than they are worth, and the railroads have done all they can to cut them out of the business. Twenty years ago, cabooses (or "waycars," as they're known in the West) were eliminated, and crews were reduced from five to two. At the same time, track controls on the high-volume coal routes are largely automated — no more picking up train orders from the train master or jumping out to switch the tracks, and no more "dark territory" (sections of track that do not have computerized signaling controlled from a central office). The entire network is operated via microwaves and satellites from a single room at BNSF's headquarters in Fort Worth. For the crew in the locomotive, it makes operating a train about as challenging as driving one of those kiddie cars at an amusement park. All the engineer has to do is stop and go with the signals and be ready to take over in case of an emergency.

Besides taking responsibility away from the train crews, automation has inspired a healthy amount of paranoia. Every BNSF engineer or conductor I talked with was conscious of the fact that each move he or she made on the train was being logged in Fort Worth. Some even believed that there were secret microphones in the locomotives that taped their conversations. It may sound crazy, but it's not altogether unjustified. In 2001, to cut down on workers' compensation claims, BNSF secretly ran genetic tests on a number of employees to see if they had a gene for carpal tunnel syndrome.

Most railroaders are far less dogmatic than management about the virtues and vices of burning coal. Perhaps this is because their politics are more liberal, or simply because hauling coal is nowhere near as fun as hauling freight. Whatever the explanation, few railroaders I talked to had any particular soft spot for coal. Anderson was even blunter about it than most. "I'd like to see them cut down on coal, even if it means losing my job," he told me, his feet up on the dash as we cruised across the prairie where his Sioux ancestors once hunted vast herds of buffalo. "I think about burning all this coal in the power plants, hurling all this stuff into the air — the problems are only going to get worse as time goes on. What about our kids, our grandchildren?"

By western standards, Crawford Hill, with an elevation of 4,400 feet, is not much more than a bump of limestone on the prairie. But having 12,500 tons of coal in tow changes the way you think about geography, and as we passed milepost 425 on our way through the bottom corner of South Dakota, the limestone bluffs that define the hill began to loom ahead of us. Crawford Hill is part of a larger outcropping known as Pine Ridge, which extends for about forty miles across the southwestern border of South Dakota and into Nebraska. Pine Ridge is the highest point in Nebraska and, until recently, the site of the only tunnel in the state. (BNSF changed the route a few years ago and abandoned the tunnel.) For railroads hauling coal out of the Powder River basin, Pine Ridge marks the final ascent and is the only significant geological obstacle between the western coalfields and the Mississippi River.

As we chugged along, Taylor pulled out his Sony handheld and showed me a track profile of the hill: beginning elevation, 3,400 feet; average grade of 1.1 percent, 800 feet in 13 miles; ruling grade, 1.72 percent; elevation, 4,400 feet at Belmont. In the graphic on the tiny screen, the hill looked like Mount Everest.

The question of the moment was whether we were going to have enough power to make it over the hill. BNSF usually attaches two helper locomotives to the rear of heavy coal trains to help boost them over the top (the helpers detach at the summit, then return to the bottom of the hill to help the next coal train make the climb). With our second engine lame, Anderson was not sure that even the two helpers would be enough. Should he ask for three? Anderson was afraid that a knuckle on one of the coal cars would break if they put too much strain on it, splitting the train in two in the middle of the hill. After a few minutes' discussion with Taylor, he decided not to risk it. He radioed Fort Worth and told them he'd need an additional helper. They agreed but told him that traffic on the hill was heavy and it would be a few hours before they could free up three locomotives.

There was nothing to do but wait. We pulled up to a stop just outside the town of Crawford, Nebraska, on a broad, open prairie. We were about a mile to the east of Fort Robinson, the historic military fort and former Sioux reservation. It was also the site of the murder of Crazy Horse in 1877 — a fact that Anderson, despite his Sioux ancestry, didn't seem terribly interested in.

If there is one thing railroaders are good at, it's waiting. They wait for track work, for traffic coming the other way, for broken trains, for weather, for signal problems. Because railroaders know they can be stranded anywhere, anytime, and are not allowed to abandon the train, every engineer and conductor starts each day loaded with food and beverages, as well as a duffle bag full of entertainment.

Railroaders are not supposed to bring along cell phones, laptops, portable DVD players, or anything else that could distract them while they're operating the train, but most conductors and engineers I met carried them anyway. Some railroaders while away the hours playing solitaire on their cell phones. Others design houses, write poetry, read novels, or look at porn.

Anderson pulled out a Bible and began reading from First Corinthians, 10:16. "I'm looking for advice on marriage and relationships," he explained. "I find it helpful."

He told me, briefly, that he had been through a hard time in the years after his brother's death. He got fat; he drank too much; his marriage broke up. "My life was a mess," he said. Then he started attending a nondenominational church in Alliance, and now he was beginning to turn his life around. "I cut my hair; I lost weight; I made new friends. Things are looking up."

I asked Anderson about the crash that killed his brother, but he changed the subject and went back to First Corinthians. A few minutes later, he put the Bible down and told me what happened. "He hit the back of another train," Anderson said coolly.

It happened at 3:30 A.M. on a stormy night near Thedford, Nebraska, a few hundred miles east of where we were sitting. Anderson's brother, Tim, ran a loaded coal train into the rear of another coal train that was stopped on the tracks in front of him. Tim's train was traveling at 42 miles per hour when it hit the other train, pushing it across another track, where it was then hit by an empty coal train that was coming the other way. Wreckage from the three trains sprawled across the tracks for several miles, causing $2.5 million in damage. Both Tim and his conductor were killed. The cab of their locomotive was so badly twisted that it took rescuers more than a day to cut their bodies out of the wreckage.

A subsequent investigation determined that the conductor was probably asleep at the time of the crash. Tim was awake, but drowsy, and passed through several "restricted speed" signals, perhaps believing the track was clear in front of him. By the time he came upon the stopped train, it was too late to do anything. The report pointed out that he had been cited twice before for passing an absolute signal, the railroad equivalent of running a red light.

Fatigue is an issue in many, if not most, railroad accidents. There are two main reasons for this. First, railroaders do not work set shifts. They are called into work as trains arrive, which means they never know exactly what time of day or night they'll be called. Although federal law requires that engineers and conductors get at least ten hours off after every shift, a lot of that time is eaten up by getting on and off the trains and waiting for shuttle transportation to take them back to the yard or to a company-owned hotel. (If a crew's shift ends in the middle of nowhere, they have to stop the train and wait for the shuttle to bring out a new crew and give them a ride back to civilization.) Railroaders' sleep habits are constantly disrupted, and their downtime is rarely long enough to ensure a good night's rest. Every engineer and conductor I met during my travels admitted as much, and most were red-eyed with fatigue.

The second reason fatigue is a problem is that locomotives are the finest sleeping machines ever built. All it takes is a dark prairie and a gently rocking train, and you're out like a light. I spent more than a few hours in BNSF locomotives listening to the conductor snore away in the wee hours of the morning. I sawed some logs myself. Most locomotives have what's called an "alerter" in the cab — a loud beeper that goes off every thirty seconds or so and that you can turn off only by punching a button. In theory, this is supposed to keep the crew awake. But as every engineer knows, there are ways of rigging the button. The fact is, despite federal laws to the contrary, a lot of sleeping goes on in the cabs of locomotives, even when — perhaps especially when — the train is hurtling across the high plains at 50 miles per hour or so.

The investigation into the Thedford crash determined that the conductor had had only four hours of sleep the day before and was known to drink twenty to twenty-five cups of coffee a day to keep awake. As the *Chicago Tribune* pointed out in 1994, rail traffic had risen 27 percent in the previous ten years, but the number of engineers and conductors was half what it had been in 1980. The *Tribune* cited a joint industry-labor review of 2 million train schedules that found crews who worked more than six trips in seven days had a higher probability of accidents, injuries, and rules violations.

Dave Anderson doesn't believe his brother was asleep at the wheel. He believes the accident was caused by a false signal — a light that should have been yellow or red but, because of a computer error, in fact was green. "I saw the tape," he said, referring to the black box log of the train's operation prior to the crash. "My brother was awake, alert, operating the horn, changing speed. The railroad says he had a signal and should have stopped. But if he was running the train properly, why would he run through a signal? Of course, the railroad claims bad signals are impossible."

Whatever the cause of the crash, it was BNSF's actions several months later that many employees will never forget. Incredibly, the railroad sued Tim Anderson's family for property damages related to the accident. In effect, BNSF, a company worth roughly $24 billion, wanted the dead engineer's family to pay for the wrecked locomotives. The Anderson family filed a countersuit, citing wrongful death. The case was eventually settled out of court. (Anderson says the family received $100,000.) Even after the settlement, however, the railroad refused to pay up in a timely manner. When Nebraska senator Bob Kerrey heard about the case, he contacted the family, offering to pressure the railroad to pay the settlement. "I then called the CEO of BNSF and told him I thought this was an outrage," Kerrey told me. "It was among the stupidest and most heartless things I've ever seen an American corporation do." Eventually, the railroad did issue a check.

I asked Anderson how this experience had changed his feelings about working on the railroad.

He snorted, then rolled his eyes. "It's made me a little more cynical, I guess," he said, as if it were the understatement of the century. Then he repeated a bit of old railroader doggerel:

> Uphill slow
> Downhill fast
> Tonnage first
> Safety last

After we had waited for more than an hour, Anderson got a call on the radio: the helper engines had arrived. Instead of three locomotives, Fort Worth had sent out four. Two engines hooked on at the back of the train and two in front, for a total of six functioning locomotives. At 4,000 horsepower each, that meant we'd have about 24,000 horsepower — roughly equivalent to the power generated by 140 Ford Explorers — pushing and pulling us up the hill. Neither Anderson nor Taylor had ever seen such an arrangement, but they were not complaining. "We're going to fly up Crawford Hill at sixty miles per hour!" Anderson crowed. With the helpers taking control of the entire train for the push up the hill, Taylor and Anderson had nothing to do but kick their feet up on the dash and enjoy the ride.

As we blasted along toward the base of Crawford Hill, diesel smoke from the hard-working locomotives in front of us billowed through the air. We rumbled through the town of Crawford, another forlorn railroad town, past stockyards with a few meandering cattle and a junkyard of rusting cars, then finally onto the piny slope of Crawford Hill.

As we climbed, I turned around to watch the broad plain of South Dakota fall away behind us. Taylor and Anderson both put their lunches away and stared out the window like tourists. Taylor became genuinely excited as the roar of the locomotives increased. After a long, straight climb, we came to the beginning of what's known as Horseshoe Curve, a broad loop of track between two hillsides that nearly doubles back on itself in the shape of a horseshoe. Taylor pointed out a spot where, for kicks, old railroaders used to jump off the train, run madly for about two hundred yards across a steep gully, and then jump back onto the train on the other side of the curve. "One guy," Taylor said, shouting above the engines, "even did it naked!"

Anderson looked down at the train's speedometer. "Nineteen miles an hour up Crawford Hill!" he yelled. "This must be a new world's record!"

As we chugged around the far side of the curve, I stepped out of the cab and into the open air on the small ledge on the side of the locomotive. I held on to the rail, wind whipping my hair, black smoke from the diesel engines swirling around me. We were running along the side of the hill now, across a spot known to railroaders as Breezy Point, because of the way the tracks just hang out on the side of the mountain. Taylor had told me that this was one of the few places in the West where you can see the entire mile-long length of a coal train as it curves around the side of the mountain. I looked back, watching the cars obediently trailing along — seventeen thousand tons of coal and iron rolling up the hill. It was glorious, an awe-inspiring feat of industrial might, even if there was something absurd about deploying so much horsepower on such a grand adventure just to haul a few thousand tons of coal to the other side of the country. But this is how America keeps the lights on.

When we reached the top, I heard victory in the engines. The roar quieted. We rolled through a cutout in the chalky mountain, passing the abandoned railroad tunnel. There was no marker at the summit. We passed an old schoolhouse and pine trees gnarled by extreme weather.

Then the ridge flattened out, and it was as if we had arrived on a giant tabletop. You could see grain silos fifty miles in the distance. It was a different geology here, the beginning of the high plains. The helpers disconnected a few miles later at Hemingford, pulling off on a sidetrack, and we were on our own again. Anderson returned to the controls, and we began a long, slow glide down toward the town of Alliance. The engines seemed relieved, still catching their breath after the hard pull. Anderson notched the brakes and held the train at a steady 25 miles per hour. I could once again feel the coal behind us, pushing us along, the tremendous weight of all that carbon rolling east. In an hour, if we were lucky, we'd be in the rail yard in Alliance. After nearly ten hours of travel, we had hauled 12,500 tons of coal exactly 121 miles. At a big power plant, this load might keep the turbines spinning for twelve hours or so. Taylor wondered aloud what his kids were doing right now, how their school day had been. Anderson looked sleepy. The sun fell toward the horizon. The steel rails stretched out before us, ribbons of progress across the prairie.

PART II

THE BURN

Chapter 5

Infinite Needs

DURING A VISIT to Plant Scherer in Macon, Georgia, one of the largest coal-fired power plants in the world, I crouched through a small opening in the wall of one of the plant's four boilers — the giant steel box where the coal usually burns in a white-hot fireball, heating the water to create the steam that spins the plant's generator — and emerged into a space that was nearly as grand as Notre Dame Cathedral. It was vast, cavernous, and ash covered; the air smelled metallic and sulfury. Thin metal tubes ran up and down the walls of the boiler, while larger ducts — they reminded me of organ pipes — ran across the ceiling above. Arcs of blue light from the welding torches of workers high up on scaffolding flashed in the darkness. Scherer's boiler happened to be shut down for a few weeks for repairs, allowing us access to the machine's innermost regions. I followed the footsteps of my guide, Danny Morton, the plant manager at Scherer. Morton worked as a boiler engineer at the beginning of his career more than twenty years ago and has never lost his enthusiasm for spelunking inside big, sooty fireboxes. As we climbed the scaffolding, he paused and ran his flashlight beam along the slag-covered wall of the boiler. "This," he said with obvious fascination, "is the belly of the beast."

This was not the first boiler I'd been inside. A few months earlier, at a coal plant in Tampa, Florida, I'd watched workers blast slag off the walls with shotguns. Two men stood at the bottom of the boiler, outfitted in goggles and hardhats. Taking aim at the crusty over-hangs, they fired away. The slag cascaded down in gray, dusty ava-lanches. "It's dirty work, mon," one of the blasters, who was Jamai-can, explained to me, "but we gotta keep it clean."

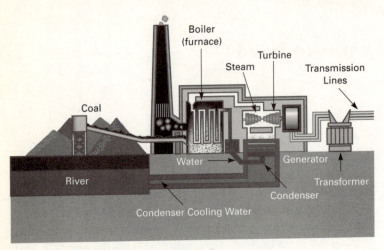

A COAL-FIRED POWER PLANT

The boiler at Scherer was twice the size of the one in Florida. As we climbed, Morton pointed out the nozzles in the corners where the pulverized coal is blown in, a swirling mix of dust and air, and the corroded spots in the pipes where the steam circulates. He talked about how the acids that are released from the coal during combustion eat away the metal in the boiler, and how subtle varia-tions in the quality of the coal change the quality of the burn. Mor-ton was clearly fascinated by every technological detail of the plant — it's his baby, after all — and told me several times about how much he misses getting his hands dirty. Now that he is plant man-ager, his job is all meetings and phone calls and spreadsheets. Lis-tening to him talk, I was reminded of the great romance of burning coal, the Cub Scout–like thrill of starting a big bonfire and watching it burn.

As we approached the top of the boiler, I tried to imagine the 200-foot-high fireball that usually blazed in this space, and the steam circulating under great pressure down to the spinning turbines and generators several floors below. I felt as if I had stepped behind the curtain in Oz to witness the dirty, clanking, primitive mechanism that creates this miraculous thing called electricity. I thought about how the electrons generated at this plant move over the wires to people's homes, bringing life to computer screens and TVs, lighting bedrooms and heating showers, and how strange and incongruous it is that all these things that are so important to our daily lives begin in this giant, slag-encrusted firebox. The more electrified our lives become, the more deeply connected we become to these big machines.

When the power goes out, there is no heat, no water, no lights. We are suddenly alone with candles and flashlights, and we realize how tenuous the things we take for granted really are, how dependent we have become on the current that flows into our houses from big coal plants like Scherer. We are hooked on the wire.

The American Psychiatric Association's diagnostic manual defines addiction as follows: "The symptoms include tolerance (a need to increase the dose to achieve the desired effect), using the drug to relieve withdrawal symptoms, unsuccessful efforts or a persistent unfulfilled desire to cut down on the drug or stop using it, and continued use of the drug despite knowing of its harm to yourself or others."

As a description of America's relationship with coal-fired power plants, this definition is fairly apt. We love them and we hate them, but we can't live without them. Between 1970 and 2000, the amount of coal America used to generate domestic electricity tripled. In some places, coal has displaced cleaner forms of generation such as natural gas and hydropower. In the 1920s, for example, Georgia Power — a subsidiary of the Southern Company, the big Atlanta-based holding company that operates Plant Scherer — generated 75 percent of its electricity at hydroelectric dams and only about 25 percent from coal plants. Today Georgia Power generates 75 percent of its electricity from coal and only 2 percent from hydro (the rest is mostly nuclear). There are some good reasons for this, including the low cost of coal and a drought in the 1920s that convinced Georgia

Power executives that it was unwise to depend on hydroelectric dams for a steady supply of electricity. But the fact remains that America's ever-increasing demand for electricity has translated into more and more coal being burned at ever bigger power plants.

And, of course, from the very beginning, we have known that this dependence on coal-fired power comes at a cost. When Thomas Edison fired up his first coal plant on Pearl Street in New York City in 1882, the residents of lower Manhattan immediately began complaining about the soot and pollution. Coal plants were eventually moved out of urban areas to the outskirts of cities and equipped with taller smokestacks to help disperse the pollutants. "The solution to pollution is dilution," an old power industry adage says. People have known for hundreds of years that too much coal smoke can kill them, but that knowledge, like the knowledge that eating too many doughnuts makes us fat, has had very little impact on our consumption. Today we know that pollution from coal plants not only causes heart attacks and triggers asthma attacks but also is helping to destabilize the climate of the entire planet. But we keep burning coal, and we keep telling ourselves that coal is getting cleaner and that whatever the downsides are to our own health and the health of the planet, they are outweighed by the benefits of the cheap electricity the coal plants provide.

Our consumption is easy to justify: cheap power is important to the American economy, the coal industry provides thousands of jobs, and whatever coal's environmental problems are, at least coal plants are not going to melt down in some radioactive nightmare or increase the risk that some Middle Eastern terrorist will get his hands on a few ounces of plutonium. But this is the logic of the beer drinker who congratulates himself for not guzzling whiskey. It is not the kind of thinking that leads us to change our lives. We like to blame others for our energy problems, but then we complain if our electric bills increase, we elect politicians who kowtow to the coal and electric power industries, and, most of all, we act as if we have no responsibility as public citizens to make intelligent choices about how we consume energy or to be held accountable for the consequences of those decisions.

All of this is true. But it is also true that America's current addiction to coal-fired electricity did not develop because we are ignorant or suffer from some character flaw. Nor was it an inevitable consequence of progress. Instead, it is the intentional legacy of a visionary Englishman named Samuel Insull.

Samuel Insull Jr. was born in London in 1859. His family was poor, troubled, and religiously devout. According to Insull's acerbic biographer Forrest McDonald, Insull's mother, Emma Short Insull, was "descended from artisans and farmers who had plodded their way through the centuries on the strength of their muscles, not their brains." His father, Samuel Sr., was an itinerant lay preacher in the Congregationalist church who was known for his rousing sermons against drunkenness and intemperance.

When Insull was six, a stroke of good fortune led to his father's being awarded a post near Oxford, and young Samuel was able to attend private school, where he proved to be an intuitive student, a natural accountant, and also good at history and politics. Insull's father expected his son to follow in his footsteps and join the ministry. Instead, at the age of fourteen, in part to help support his family, Insull decided on a career in business.

His first job was as a secretary and bookkeeper at an auction house in London. Two years later, largely on the strength of his excellent shorthand, he won a night job as a stenographer for Thomas Bowles, the legendary editor and founder of *Vanity Fair*. Through Bowles, Insull was exposed to the ideas of many of the most influential figures of the day, including P. T. Barnum and, more important, Thomas Edison. "Reading about [Edison] and his work," Insull later wrote, "and the romance of his life — from train boy and telegraph operator to an internationally known inventor — naturally aroused my youthful enthusiasm." Not long after that, Insull was fired from his job at the auction house to make way for one of the owner's sons. He thought his life was over.

Instead, it had just begun. Combing through the classifieds in a London newspaper one day, Insull answered an ad for a stenographer placed by an American banker named George A. Gouraud. After Insull took the job, he learned that Gouraud was one of Edison's agents in England; he was thrilled. Before long, Insull found himself on a boat bound for New York. Waiting for him in Manhattan was a job as the personal secretary to the most famous inventor in the world.

It was Edison, of course, who had figured out a way to use coal to generate electric light. On September 4, 1882, on a narrow, crowded street near the docks along the East River in lower Manhattan, Edison flipped a switch that fired up the world's first coal-fired electric power plant, or dynamo, as it was called in those days. Nicknamed "Long-Legged Mary Ann" by some of the workers, for the way the tall cylinders of the plant resembled the legs of a woman, Edison's dynamo lit up several blocks of lower Manhattan, including the offices of the *New York Times* and of financier J. P. Morgan, one of the richest men in America and one of Edison's main financial backers.

Long-Legged Mary Ann was a dirty, loud, ungainly beast. Coal arrived via metal doors in the sidewalk, where it rumbled down a chute into the cellar. Conveyors then carried it up one floor, where filthy, soot-covered workers shoveled the coal into one of four Babcock and Wilcox boilers, which fed the heat into the six steam engines one floor above. The shaft of each engine was coupled directly to a generator via leather straps. Each generator stood six feet tall and produced enough current for 1,200 lamps. Black smoke and soot exited through a couple of makeshift chimneys in the roof, blackening the entire neighborhood.

It's hard to grasp the importance of Edison's accomplishment today, when electric light is so commonplace, but building a system to harness the heat latent in a ton of coal and transform it into light ranks among the greatest technological feats of all time. In less than three years, Edison had not only designed and built the first dynamo and the first workable incandescent bulb, but he also had created the first electric power system — the wires, transformers, and plugs — and used his own money and political clout to get it built in downtown Manhattan.

Edison's achievement opened many people's eyes, but it also raised many questions. One was where the coal was going to come from to feed all the dynamos Edison aspired to build around the country. And even if America had plenty of coal, was it really wise to burn it for electricity? Nikola Tesla, the eccentric genius and rival of Edison who designed, among other things, the first electric motor, believed that nonrenewable resources such as coal needed to be conserved. Windmills, he declared, should be placed on roofs immediately. (Tesla was also an early advocate of solar and hydro power.) Edison dismissed such concerns, believing that shortages of coal and other natural resources could be put off for "more than 50,000 years." The forests of South America alone, he argued, would provide fuel for that long.

A more urgent issue for Edison and his bankers was how best to capitalize on his invention. Edison wanted to keep a grip on the entire system, creating what amounted to a vertical monopoly: building the power plants, laying the wires, supplying the light bulbs. He would make money by selling the current, similar to the way gas companies made their profit by selling gas. J. P. Morgan had a different idea. Instead of creating a commodity (electricity), Morgan wanted to structure the electricity business around selling the machinery to make the power. To Morgan, it was simpler to build a widget and sell it at a profit than to get involved in the whole complicated business of creating and managing electricity. In fact, Morgan's vision looks a lot like what is known today as distributed generation — lots of small generators located close to consumers.

If Morgan's vision had prevailed, the structure of the electric power industry might look very different today. Instead of pulling in electricity over wires from distant power plants as we do now, we might be generating electricity much the way we generate heat: with a small device in the basement. In the long run, this might have been a much more efficient way to do things, because it might have allowed heat and electric power generation to be combined into a single unit: the heat created by the generator could have been reused to, say, warm a house or office building. But Morgan's model of the industry had at least two things going against it. One was that the early dynamos were noisy, dirty machines that people didn't like to be around — far better to locate them out of sight and out of mind. The second was that Edison himself was a bit of a control freak. He understood better than anyone how complex and fragile this new electrical system was, and he knew that if it was going to have a chance of succeeding, he needed to remain the master of it.

Beyond this, Edison's invention raised broader questions, such as what, exactly, was electricity good for? Incandescent light was nice, but it was very expensive to produce, and most cities in America were already well lit by coal gas. Many Americans were not sure what to make of electricity. They saw it as a mysterious new force, associated both with danger and electrocution and with spirituality and vitality. Like all new technologies, electric power needed its "killer app" — the one application that would make it irresistible to consumers. And no one, not even Edison, had any idea what that might be.

Then young Samuel Insull arrived from across the Atlantic.

Edison and Insull were an odd pair — Edison a disorganized, half-deaf American genius; Insull a twenty-one-year-old, clean-living Brit with a cockney accent. Insull arrived just as Edison was putting the finishing touches on Long-Legged Mary Ann, and he was involved in every detail of the business, from securing a steady supply of coal from Pennsylvania to taking care of Edison's personal investments. Insull proved to be a quick study, with a mind for detail and a shrewd political instinct, as well as an indefatigable worker. Within a few years, Edison put Insull in charge of Edison Electric's manufacturing operations and told him to move the whole thing to Schenectady, New York. "Do it big, Sammy," Edison told him. "Make it either a big success or a big failure." Over the next six years, Insull built up Edison's Schenectady plant from two hundred to six thousand employees and made the city the manufacturing capital of the electricity industry.

When Charles Coffin, a tough-as-nails former Massachusetts shoe manufacturer, acquired Edison Electric in 1892 and merged it with Thomson-Houston, a rival company that Coffin controlled, he called the new company General Electric and offered Insull a job as vice president. Insull, like Edison, despised Coffin. Besides, Insull wanted to run his own show. He lobbied for and won the job as head of Chicago Edison, a licensee of Edison's technology and one of the largest and fastest-growing power companies in the country. More important, it was at the epicenter of America's coal-fired industrial heartland — an ideal place for Insull to begin reinventing the electric power industry.

As Insull knew better than anyone, one of the problems with Edison's vision of the industry was that it was extremely capital-intensive. The electric power business was a lot like the railroad business: high initial capital costs meant high barriers to entry, which, from a monopolist's point of view, was ideal, because it kept competitors out. In comparison to the railroads, however, financiers were generally more willing to provide the capital to build power plants than railroads, provided returns were secure. But if too many power plants were built at once, everyone would go broke. So what was a good capitalist to do?

In 1892, barely a year after he took over Chicago Edison, Insull stunned a gathering of power industry executives by telling them that he believed that competition among power companies was "economically wrong," largely because the cost of building duplicate wires and power plants was economically untenable. In fact, Insull argued, power companies were "natural monopolies" that should be regulated by the state. Given the predatory instincts of nineteenth-century industrialists, state regulation was a radical idea indeed.

It took Insull's peers a while to see that submitting to regulation was a brilliant solution to the industry's problems. By the time Insull proposed this idea, he had done enough business with the notoriously corrupt Chicago political machine to understand that the "regulators" he was proposing would merely go through the motions of overseeing electric power utilities. Any regulator who got too pushy could be (and was) quickly disposed of. By accepting the yoke of state regulation, Insull killed the drive for municipally owned power systems that had been growing in many cities around the country and that was the biggest challenge to the industry. It also gave power pioneers a huge advantage, cutting out the threat posed by entrepreneurial upstarts, who had no hope of competing against these "natural monopolies," which enjoyed special legal privileges and protections. And the power pioneers could achieve all this while pretending to work in the interest of the people. "No shrewder piece of political humbuggery and downright fraud has ever been placed upon the statute books," Milwaukee mayor Daniel Hoan said in 1907. "It's supposed to be legislation for the people. In fact, it's legislation for the power oligarchy."

Important as this move was, it wasn't the end of Insull's genius.

To most operators, establishing a monopoly was desirable because it would enable them to charge higher prices and thus make more money. But Insull was interested in establishing a monopoly for the opposite reason: instead of raising prices, he would use his monopoly power to *cut* them.

From the beginning, Insull's interest in big, centralized power plants was driven by an essential economic fact: power plants are expensive to build but comparatively cheap to run. One of the principal mysteries of the business is that the total expenses of a power plant are about the same no matter how much electricity is used. (The cost of coal is minor compared to the cost of building and maintaining the plant and the surrounding power network.) In this way, big power plants are like big airplanes: because the cost of flying a 757 from, say, New York to Los Angeles is fixed, the airline makes far more profit on the last ten seats that it sells than on the first ten. Similarly, Insull realized that he was far better off charging a thousand customers ten cents per kilowatt-hour than charging a hundred customers one dollar per kilowatt-hour. Furthermore, Insull reasoned that if prices were low enough, manufacturers could be easily convinced to retire their own power plants and sign up with one of Insull's companies; homeowners would be willing to let Insull install electrical outlets, which in turn would create markets for things such as electric irons, refrigerators, and other household appliances. In effect, cheap power stimulated demand. And once people got hooked, Insull understood, there was no going back.

Cheap power had enormous social and economic benefits, of course. It opened the door to new technologies from refrigerators to radios and enabled a revolution in manufacturing, which lowered the prices of all kinds of consumer goods. As early as 1914, nearly every type of electric appliance known to us today could be obtained: not just electric stoves and heaters, but electric kettles, toasters, dishwashers, washing machines, vacuum cleaners, mixers, potato peelers, knife grinders, and vibrators (a very popular item, usually marketed as a way to relieve female "hysteria"). But these items were expensive, and it wasn't altogether clear to many people why they needed an electric toothbrush or teakettle. After all, what was wrong with the old way of doing things? Electricity required not just new technology but also new marketing skills. And Insull turned out to be as good at figuring out new ways to promote electricity as he was at figuring out new ways to generate it.

From the beginning, the rise of electricity was connected with modernism and social progress. After the Chicago World's Fair in 1893, where 22 million fairgoers got a glimpse of the futuristic all-electric Dream City, electricity became an instant icon. In cities across America, from Akron to San Francisco, local citizens banded together to bring electric lights to their Main Streets. More than anything else, it was simply a way of showing how up-to-date they were.

But stimulating demand for consumer goods was more difficult. Companies such as General Electric launched huge print advertising campaigns that sought to instill an "electrical consciousness" in the public. As one GE publicity man explained, "It was not until people were told, emphatically and repeatedly, what the telegraph, the telephone, and the electric light could do for them, not until they had seen the possibilities of these strange new devices demonstrated before their very eyes, that they began to develop a want."

The best way to develop "a want," the power companies found, was to promote electricity as clean and sanitary: no more soot from the coal stove! As one 1927 ad put it, "For Health's Sake — Use Electricity." Increasingly, cleanliness became equated with good health, modernity, and social progress.

No one was better at exploiting these connections than Insull. He understood how to create a want for power by connecting it to the deepest human emotions, including motherhood and the desire for immortality: "How long should a wife live?" asked one typically sharp Insull company newspaper ad. "The home of the future will lay all of its tiresome, routine burdens on the shoulders of electrical machines, freeing mothers for their real work, which is motherhood. The mothers of the future will live to a good old age and keep their youth and beauty until the end." Insull was also a pioneer in advertising to children. Recognizing that kids would become "the customers, the investors, the voters, and the lawmakers of the future," he provided schools in the Chicago area with a thirty-two-page color brochure entitled *The Ohm Queen* that touted the wonders of electricity.

As a businessman, Insull knew that the deployment of cheap power was of enormous strategic benefit. Whatever question or problem he confronted, especially if it was from reform-minded politicians worried about his increasing stranglehold on power generation in the Midwest, his response was always the same: cut rates. More often than not, the problem — or the politician — went away. Like Sam Walton fifty years later, Insull discovered that as long as he kept prices low, he could get away with pretty much anything.

The key to this entire strategy, Insull learned, was building ever bigger and more efficient power plants. The bigger the plant, the easier it was to leverage efficiencies in distribution and transmission. Insull referred to this as "massing production," and his PR team later shortened the phrase to "mass production." Like Edison, however, Insull faced a number of technical problems when he tried to increase the size of his plants. Limitations on the distance of power delivery were simple enough to overcome. By the turn of the century, alternating current had won out over direct current, making it much easier to send power over long distances, and improvements in transformers made it easier to control the voltage. The more troublesome limitations had to do with the big, clanking, piston-driven steam engines that were used to power generators. Wasn't there a better way to do this?

In 1902, Insull heard about an English engineer named Sir Charles Parsons, who had built a small steam turbine for racing boats. The turbine was a radical simplification of the old piston-driven steam technology, replacing belts, crankshafts, and pulleys with a single, elegant fan. Insull asked his engineers if this turbine could be adapted for electric power plants. No way, they told him. Undeterred, he contracted with his old nemesis at General Electric, Charles Coffin, to build him one anyway, agreeing to share the cost if it turned out to be a spectacular failure.

Seventeen months later, Insull found himself standing in front of a brand-new power plant on Fisk Street in Chicago. It was like nothing anyone had seen before — sleek, small, and tremendously efficient at transforming coal into electricity. It was essentially a jet engine pointed toward the sky. More important, it was scalable; you could build it as big as you wanted.

Within a few years, the old steam engines were dead, and the era of the big coal plant had begun. The triumph of large coal-fired turbines was a boon for cheap power. The average price Americans paid for electric power fell from $4.53 per kilowatt-hour in 1892 to 62 cents in 1927 to 47 cents in 1937. And as the price fell, consumption grew. Between the 1920s and the 1970s, the demand for electricity roughly doubled every decade.

Insull's empire expanded with each kilowatt consumed. In 1907, when he renamed his Chicago-based utility Commonwealth Edison, it was already sixty times larger than it had been when he arrived in the city just fifteen years earlier. By 1916, Insull controlled 118 power systems operating in 9 states, all organized into subregional and regional holding companies. Insull's main holding company, Middle West Utilities, was at the top of the pyramid. By the 1920s, Insull was a very rich man, with a personal fortune of $150 million, a palatial house on Chicago's Gold Coast, and a 4,300-acre farm in Libertyville, Illinois.

Together, Edison and Insull created a nation of electricity junkies. In 1905, less than 10 percent of homes in America were wired. By the late 1920s, 75 percent were, including almost all urban dwellings. By then, electric irons, hair dryers, telephones, phonographs, and radios were commonplace. Coal stoves were gone, replaced by electric or natural gas heat. Electric light and air conditioning allowed the development of department stores such as Marshall Field's in Chicago and John Wanamaker's in Philadelphia, which were soon stocked with items — from fancy perfumes to electric teakettles to imported clothes — that had been impossible to find in the pre-electric era. America's consumer culture was born.

Edison's and Insull's innovations inspired another key development of the twentieth century: the automobile. Young Henry Ford, whose formal education was limited but who was handy with machinery, went to work for one of the Edison company's early licensees, Detroit Edison, advancing from machine-shop apprentice to chief engineer. He quit in 1899 to start an automobile company. At the time, automobiles were still novelty items for the rich and do-it-yourself engineers. But Ford's understanding of electricity, as well as the availability of cheap power to run electric motors and other devices, was fundamental to the development of the assembly line he used to build the Model T, which began production in 1908 and revolutionized American industry.

In the 1920s, as the demand for electricity skyrocketed, a merger frenzy swept through the electric power industry. In the space of a few years, small regional utilities were transformed into corporate giants. In the South, Georgia Railway and Power Company, the Atlanta-based utility that had been started by a Yankee blue blood named Henry Atkinson, was subsumed by a larger holding company, Southeastern Power & Light. In 1927, Southeastern Power & Light became part of an even bigger holding company, Commonwealth & Southern, which was headed by Wendell Willkie, a Wall Street lawyer who became Franklin D. Roosevelt's Republican opponent in the 1940 presidential election. Commonwealth & Southern eventually gained control of electric power subsidiaries in ten states — Michigan, Illinois, Indiana, Ohio, Pennsylvania, Mississippi, Alabama, Florida, Georgia, and South Carolina — as well as eight transportation companies, an ice company, three land companies, and a water company. In New York, another empire builder, Sidney Mitchell, snapped up small electric companies in Indiana, Kentucky, Michigan, Virginia, and West Virginia. These holdings later became the basis for American Electric Power (AEP), the largest power company in the nation today.

In terms of industry clout, however, nothing rivaled Insull's Middle West Utilities, which, at its peak, was worth more than $3 billion and controlled utilities that generated one eighth of the electricity in the United States. The company was a mountain of holding companies and sub-holding companies, leveraged every which way and tangled in a financial web that was so complex that few historians have managed to unravel it. At one point, Insull himself held sixty-five chairmanships and eighty-five directorships and was president of eleven corporations.

In the end, Insull and his imitators were too successful for their own good. The power trusts had acquired a monopoly not just on electricity, but on public life, which by the late 1920s led to abusive pricing in some areas and near-total control over the economic well-being of the country. As for the state regulators who were supposed to keep the power trust in check, Senator George Norris of Nebraska, one of the most ardent power industry foes, argued that they "can no more contest with the giant octopus [of the power industry] than a fly can interfere with the onward march of an elephant." Not surprisingly, reformers like Norris began to call for public ownership of power companies. The industry fought back with harassing litigation, rate wars, and a massive propaganda campaign, including smearing public power advocates as "un-American" and tied to "Bolshevik" ideas. In 1928, an investigation of the Federal Trade Commission uncovered examples of bribery, payoffs, and economic blackmail that the power industry used to persuade academics and newspaper editors to support their position. President Franklin Roosevelt later denounced it as "a systematic, subtle, deliberate and unprincipled campaign of misinformation and propaganda, and if I may use the words — of lies and falsehoods."

Soon after Wall Street crashed in 1929, Insull's highly leveraged empire collapsed, taking several billion dollars in shareholders' money with it. (Stock in Insull's company was widely held by everyday investors, so-called widows and orphans, many of whom lost a good percentage of their life savings.) Vilified in the press and indicted on charges of embezzlement, mail fraud, and evading the federal bankruptcy act, Insull fled to Europe. Unfairly or not, he became a symbol of corporate greed and the excesses of the era. "The Insull failure," President Franklin Roosevelt declared, "has opened our eyes. It shows that . . , the development of these fraudulent monstrosities was such as to compel ultimate ruin; that practices had been indulged in that suggested the old days of railroad wildcatting; that private manipulation had outsmarted the slow-moving power of government." In 1934, Insull was arrested in Turkey and extradited to the United States. During an eight-week trial, Insull's lawyers portrayed him as a Horatio Alger, a hard-working kid who rose from poverty, only to be duped by rich, clever bankers. He was acquitted of all charges and died of a heart attack in a Paris subway in 1938.

Today Insull is remembered simply — and unfairly — as a crook. But his real legacy, according to social historian Harold Platt, is his creation of "the gospel of consumption." Insull, Platt argues, "set in motion a self-perpetuating cycle of rising use and declining rates that eventually enveloped an urban society in a ubiquitous world of energy."

Most of us still live in Samuel Insull's world. We get our electricity from big, centralized power plants, we pay a flat rate for cheap power, and we are constantly plugging new devices and appliances into the wall. Cell phones, DVD players, laptops, engine block heaters, heat lamps for pet snakes, Christmas tree lights, portable air conditioners, humidifiers, dehumidifiers — we just plug them in and never think about how much power they suck up. Because we don't see any soot raining down from the sky, most of us still think of electricity as a clean, modern, and inexhaustible resource. And the power companies are still doing their best to promote consumption, even if they temper it occasionally by paying lip service to conservation and efficiency. After all, they are in the business of selling electricity: the more you use, the bigger they grow. In the past century, that dynamic hasn't really changed.

In 1935, largely in response to the excesses of Insull, Congress passed the Public Utility Holding Company Act, which broke up the big electric power companies. The new law limited the geographic area that utilities could operate in and restricted the kinds of business they were allowed to invest in. In addition, electricity rates were now set by regulators using a method called "cost-plus pricing," which guaranteed utilities a fixed return on their investments. In return, the utilities were required to serve all customers in their local areas, and they were forbidden to compete with one another regarding price, quality, or services.

For a while, it seemed like a good remedy. Politicians got the stable and secure supply of electricity they sought, while power companies got their "natural monopolies" further codified. With no competition, gentle oversight, and tremendous political leverage, the power companies felt no pressure to innovate or change the way they did business. From the 1920s to the 1970s, the electric grid in

America grew essentially by replication — more wires and bigger power plants. The entrepreneurial thrill of the Insull years ossified into the tedium of managing centralized bureaucracies. In the power industry, this era is inevitably referred to as "the golden years."

By the 1970s, it was clear that the reforms of the 1930s had created their own set of problems. One was technological stagnation. Power plants and the electric grid had gotten bigger, but they weren't getting any more efficient. Another problem was the runaway cost structure of building ever larger power plants, especially nuclear plants, which often went billions of dollars over budget and nearly bankrupted the utilities that built them. As a result, instead of the average rates for power declining as they had between the 1920s and 1950s, they started to climb, rising by 60 percent between 1969 and 1984. The combination of bloated costs and irate customers proved that when it comes to stifling progress, a "natural monopoly" does the job about as well as any other kind of monopoly.

To power industry reformers, giant coal plants such as Scherer (construction of the plant began in 1975, at the height of the industry's troubles) were clear evidence of just how bloated and backward the industry had become. Back in the 1920s and 1930s, when Insull was building the first big coal burners, he had moved them out of urban areas to reduce pollution in cities and allow him to build ever larger plants. Efficiency-wise, moving power plants to remote locations meant that 10 to 15 percent of the power was lost during transmission to users. Power plants that had been located in cities also saved energy by reusing the steam to heat nearby buildings, often eliminating the need for separate heating systems. At remotely located plants, waste heat was simply vented into the air. At big coal plants today, as much as two thirds of the energy they produce is simply dumped into lakes, rivers, and cooling towers.

By the 1970s, it was clear to many engineers that there was a better way. New technology — particularly small, mass-produced turbines powered by natural gas — made it far more efficient to locate power generators close to consumers, which would eliminate the loss of energy due to transmission. These gas-fired power plants also were much cleaner than coal plants, so pollution was not as much of an issue. Even better, by locating the plants closer to customers, the waste steam from the plants could be used for home or industrial heating. All in all, it amounted to what should have been a revolution in the electric power industry, one that would lead to a much more decentralized, locally controlled, and creative electric grid.

But the power industry wasn't interested. "The industry had ossified," wrote Thomas Casten, the CEO of an early independent power company who is known for his progressive views. The heart of the problem, Casten argued, was that, as regulated monopolies, utilities were allowed to set electric rates at a price that guaranteed them a fair return on their capital investment. To prevent excessive profits, the utilities were required to pass 100 percent of any efficiency gains on to customers. This meant that utilities had no financial incentive to adapt new, cheaper, smaller, more efficient generation technologies. "In fact, such local generation erodes the rationale for continued monopoly protection," Casten wrote. "If one can make cheap power at every factory or high rise apartment house, why should society limit competition?"

In the 1980s and early 1990s, deregulation began to shake up the pampered electricity business. The experiences of airlines, trucking businesses, and telephone companies were demonstrating that deregulation brought generally lower prices and more innovative services to customers. But the electric power industry — particularly in the regions dominated by big coal burners — fought hard against this change, usually by citing the low cost per kilowatt-hour of energy produced at big power plants, while ignoring the other hidden costs and inefficiencies in the system, such as the fact that these plants often cost a billion dollars or more to construct. As a result, the restructuring of the electric power industry has been a slow and tortured process, with a number of high-profile disasters, such as the rolling blackouts in California in 2001 and the collapse of Enron, a vocal supporter of restructuring, shortly thereafter. Today the battle continues, with states such as Texas, Pennsylvania, and Illinois moving toward more open markets, and other states such as Georgia and Wisconsin remaining committed to the conventional regulatory model.

There's no doubt which side of the restructuring debate Big Coal is on. Coal is unabashedly aligned with the old paradigm of big, centralized generating plants. This is partly an alliance of necessity: moving huge quantities of coal from the mines to the power plants — Plant Scherer consumes more than 34,000 tons a day — is difficult and expensive, which means that burning coal is economical only if a plant is located near a river or railroad. This in turn usually means that the plant is poorly suited for local generation. And, of course, the pollution from a coal plant makes it an unwelcome neighbor. (Even the cleanest new coal plant is significantly dirtier than a gas-fired plant.) Another factor is the cost of construction. Like a nuclear plant, a coal plant is very expensive to build but relatively cheap to operate. For independent power producers, which have to finance the plants themselves, this is a major hurdle. For regulated utilities, whose profits are calculated as a percentage of capital invested, the high cost of building a coal plant is actually a virtue.

But the real obstacle to change is what some people in the industry affectionately call "the big dirties." Simply put, these older coal plants — most of them built in the 1960s and 1970s, before pollution controls were mandated — produce electricity so cheaply that it is virtually impossible for other power plants to compete with them. When I visited Plant Scherer in 2003 and 2004, for example, it was generating electricity for about $20 per megawatt-hour — about half the price of competing natural gas plants. (A megawatt is equal to 1 million watts. Ten thousand 100-watt light bulbs burning for one hour consume one megawatt-hour of electricity.) And Scherer is nowhere near the cheapest (or the dirtiest) plant. Some old coal burners in the Midwest generate power for as little as $8 per megawatt-hour. In regulated markets, these cheap prices are passed on to ratepayers. But in restructured or partly restructured markets, these plants can sell their excess power to regional wholesale markets, often making whopping profits of $25 per megawatt-hour or more, while other, cleaner power plants are counting their profits in nickels and dimes. More important, the construction costs of many of the big dirties have long since been paid off, meaning that every dollar they take in (above fuel and operating costs) is pure profit. "They are legal mints," Dale Simbeck, a power industry consultant, told me.

The profits that the old coal plants without pollution controls rake in are not simply a consequence of the fact that coal is cheap and plentiful. These plants are legal mints because the regulatory system fails to account for the social, environmental, and public health costs of burning dirty coal — what economists call "externalities." All other things being equal, dirty megawatts from old coal burners are almost always cheaper than clean megawatts from other sources. The real costs of those dirty megawatts — the devastated mountains of West Virginia, the heart attacks and asthma caused by air pollution, the pumping of greenhouse gases into the atmosphere — are all offloaded onto the public. How would competition in the electricity markets change if these costs were factored in? According to a ten-year study known as ExternE, which Princeton University professor and energy expert Robert Williams calls "state of the art," factoring in just the public health effects of air pollution from U.S. coal plants would add an average of about $13 per megawatt-hour to the price of coal-fired power. (This does not include damages connected to mining, nor does it include costs related to global warming.) For the big dirties, added costs could be as high as $33 per megawatt-hour. In comparison, the cost of externalities on a natural gas plant are only about 40 cents per megawatt-hour. In a market that accurately reflected the true cost of power, old coal burners would be shut down because the price of the power they generated would be too high for the market to bear.

For electricity, price is really the heart of the problem. Not only are blasted mountains, dead coal miners, and medical bills for asthmatic children not reflected in the way electricity is priced, but neither is the real cost of generating it. In many regions of the country, residential customers still pay a single flat rate for electricity that is set by state regulators — the ghost of Insull again! — and has little or nothing to do with the actual price of generating that power at the time the customer is using it. This means that at 5:00 P.M. on a hot summer day, when the air conditioners are cranking, the power load is high, and every power plant on the system is running, it might cost the local power company $100 to generate or buy a megawatt of electricity — yet electricity customers are still paying the same flat rate they pay in off-hours, when the price to generate electricity might be only $20. As customers, we are totally disconnected from the true price of power. And the message we get from the power company is this: *Don't worry, be happy, leave the lights on. If there's trouble, we'll take care of it.*

It doesn't have to be this way. One can easily imagine an electricity market that is structured more like the market for hotel rooms or airline travel — when demand is high, prices rise; when demand is low, they fall. People could compensate for higher prices during times of high demand by switching off lights, avoiding doing laundry, and turning off air conditioners. This kind of dynamic pricing might inspire new devices to help people monitor their power consumption, such as easy-to-read meters located inside the house or smart chips programmed into air conditioners and other appliances that can be programmed to shut down when electricity prices reach a certain level. And why does the transmission grid have to be one-way? Why can't neighborhoods that want to band together and buy solar panels or a small gas turbine upload their power onto the grid? Why can't the grid be open instead of closed, democratic instead of authoritarian? "What we need is a system where entrepreneurial types can come in and compete and try out different things," Vernon Smith, the winner of the 2002 Nobel Prize for Economics, has argued. "Some will lose money and go broke. Some will hit it big. That's what's happening in telephones, and it's not happening in electric power."

In some regions of the country, Insull's vision of a centralized power structure is beginning to break down. But it is happening slowly, and it is happening most slowly in states that burn a lot of coal. Coal plants are, after all, the embodiment of centralized control and monopoly power. They were built for a nation hooked on the gospel of consumption and protected from the true costs of their habit by distorted markets and feckless or corrupt regulators. As that old paradigm of the industry fades, coal will undoubtedly have a harder time. The last thing any big coal burner wants is a fluid, transparent, flexible market; an open transmission grid with easy interconnections; and a population of knowledgeable consumers.

Ironically, if Edison and Insull were alive today, they would probably not stand arm in arm with Big Coal, lobbying for the status quo. After all, Edison and Insull changed the world by inventing new ways of doing things. Like today's Silicon Valley entrepreneurs, they understood that technological change is something to be welcomed, not feared. And they believed in the power of creative destruction. "In my business," Insull once said, "the best asset is a first-class junk pile."

Chapter 6

The Big Dirty

MORE THAN 2 MILLION death certificates were filled out by doctors and health officials in the United States in 2004, and in the space allotted for cause of death, it's a good bet that not one of those certificates read "pollution caused by coal-fired power plant." Fifty years ago, in industrial states such as Pennsylvania and Ohio, people were still dropping dead in the streets on days when air pollution was particularly bad. In China and India, they still are. But in many parts of America today, coal plants are much cleaner than they used to be, the smokestacks are higher, and the effects of pollution are more widely dispersed. The fact that most Americans no longer fear that pollution from a coal-fired power plant will kill them is both a sign of progress and a dangerous illusion. Pollution from coal, both the mining and burning of it, is still a major public health issue in the United States. Tens of thousands of people are still dying prematurely each year as a result of America's romance with coal. But now it happens in slow motion and in ways that don't translate easily to death certificates.

One place where people understand this all too well is Masontown, Pennsylvania. Masontown is the home of one of the dirtiest coal plants in America — Hatfield's Ferry, a 1,710-megawatt coal burner that is owned and operated by Allegheny Energy. When construction was completed on Hatfield's Ferry in 1971, it was a typical coal burner for the time and built without modern pollution controls of any sort. (When the plant was designed, they weren't required.) In southwestern Pennsylvania circa 1971, especially in an old coal mining community like Masontown, no one gave much thought to the health effects of dirty air. Pennsylvanians had been living with soot under their noses for one hundred years. They not only accepted it; they took pride in it, thinking of it as a kind of industrial badge of honor.

Not long after the plant opened, a young couple named Charlotte and Donald O'Rourke bought some land that used to be part of an old farm and built a house within sight of the smokestacks. Sometimes the soot settled over their house like black snow, and the air was often so thick and hazy that it was hard to breathe. But by then Charlotte was a busy mother of two kids, working part-time as a tax accountant, and she just accepted the haze and soot as part of everyday life. Donald, who drove a dairy truck for a living, never complained about it either. "We thought of the coal plant as more of a nuisance than anything," Charlotte says now. "It was no fun to always wipe the soot off our cars and windows, but nobody thought it was dangerous or harmful."

In fact, Charlotte and Donald, like most residents of Masontown, saw Hatfield's Ferry as the engine of the local economy: as long as the smoke was pouring out, some factory somewhere was humming and people were working. But in the 1980s, when the collapse of American manufacturing began, the coal plant didn't save them from anything. Western Pennsylvania was among the hardest-hit regions of the country. The coal mines closed, the factories cut back, and communities like Masontown became desperate places, marked by abandoned schools and thriving funeral parlors.

Then, in 1989, Donald developed a pain in his back. Charlotte wasn't worried at first because he was only fifty-six years old and rarely sick. He went to the doctor and was diagnosed with renal cell carcinoma, a deadly form of kidney cancer. Within six months, he was paralyzed and in severe pain. When Charlotte visited him in the cancer ward at the hospital in Pittsburgh, she was stunned to find three other people from the area around Masontown — a town with a population of only 3,500 — who were dying, too. How could that be? When she searched for explanations, she immediately thought of the power plant, which was by far the largest polluter in the area, and the soot that had rained down on them for almost two decades. "Could pollution have caused this?" she asked Donald's doctor. He told her it was certainly possible, but the only way to prove it was to find a pattern of disease in the area, which would take years of study. A few months later, Donald died.

Today, Charlotte still lives in the same house, and the smoke and soot are still pouring out of Hatfield's Ferry. She is a white-haired lady of seventy-two now, and she does not mince words about what has happened to her world. In the past few years, she has learned a lot about the power plant. She knows that it is one of the dirtiest plants in America; that the company that runs it today, Allegheny Energy, has been sued by five states for failing to install modern pollution controls on its old coal plants; and that while many people in her town believe that the plant is killing them, few are willing to speak out about it because if the plant closed down, it would be devastating to the local economy — or what's left of it. Greene County is now the poorest county in Pennsylvania. Charlotte also has learned about the strange ways people become dependent on the very things that might kill them. She knows that she should move away, but she loves this place — loves her friends, the hills, and the house that her husband built with his own hands. "I can't leave," she says firmly. "This is my home. This is where I grew up."

But her life is different now. She keeps her windows closed tight year-round. Even so, the soot works its way in, and she wipes her windowsills down nearly every day. She doesn't garden anymore, because that means spending time outside and breathing the air, which she tries not to do. (When her grandchildren visit, they stay inside, too.) When she looks out the living room window, she sees the old white farmhouse where her friend Ida, who died in 2004 from bladder cancer, used to live. She talks almost daily to her friend Gloria, who lives down the road and has breast cancer and asthma. Charlotte's brother-in-law, who is fifty years old, attaches himself nearly every day to a breathing machine. The young son of a hairstylist in town has aplastic anemia, a rare and serious blood disease that has been linked to exposure to toxic substances. At the elementary school, teachers keep inhalers handy for kids with asthma. Few people eat the fish taken from the local rivers, including the Monongahela, in part because coal plants in Pennsylvania are among the largest emitters of mercury in the nation.

When Charlotte sits in her living room, she can look out over the rooftops of the town — a poor, sick, dirty place that barely resembles the town where she grew up. The raspberry bushes don't grow on the hill above her anymore, and just about everyone she knows drinks bottled water and is afraid to go outside. Still, the soot keeps falling, the cemetery up on the hill behind her house keeps expanding, and nobody — not the mayor, not the people she's talked to from the Pennsylvania Department of Environmental Protection or the Department of Health, and certainly not the president of the United States — seems to give a damn.

"You really don't have to be a scientist to see what's going on around here," Charlotte told me one afternoon as we sat in her impeccably clean living room. "We live under the plume, and people are sick and people are dying. I mean, how complicated is it, really?"

Just a few months after our visit, Charlotte was diagnosed with precancerous cells in her esophagus.

It is indisputably true that, in many parts of the country, the skies are bluer today than they have been in one hundred years. Thanks mostly to tough environmental laws that have forced Big Coal to clean up its act, new pollution controls mean that coal is burning more cleanly than ever. As various coal industry associations and spokespeople never tire of pointing out, since the Clean Air Act was passed in 1970, the average emissions rate (total emissions per unit of energy produced) from coal plants in America has dropped by 77 percent for sulfur dioxide, 60 percent for nitrogen dioxide, and 96 percent for large particles of soot. This is an important achievement, and one that the industry and the environmental community are justifiably proud of.

But it's also important to put this progress in perspective. By any measure, the volume of pollution released by coal plants remains staggering. Nationwide, power plants account for two thirds of all sulfur dioxide, 22 percent of all nitrogen oxides, nearly 40 percent of carbon dioxide, and a third of all mercury emissions.

Coal plants also release some sixty varieties of what the EPA terms "hazardous air pollutants," including known toxins such as lead (176,000 pounds), chromium (161,000 pounds), arsenic (100,000 pounds), and mercury (96,000 pounds). And that's just what goes into the air. Each year, coal plants produce about 130 million tons of solid waste — about three times as much as all the municipal garbage in the nation. This combustion waste — fly ash, bottom ash, scrubber sludge — is laced with heavy metals and other potentially toxic compounds and is routinely pumped into abandoned mines or impoundment ponds, where, if it is not handled correctly, it can leach into aquifers and water supplies.

It is also true that despite the broad trend toward cleaner air, progress is not equal everywhere. According to the EPA, half of all Americans still live in areas where air pollution levels exceed national health standards. In the Northeast, for example, the level of small-particle pollution — tiny bits of soot, acid droplets, and toxic metals that have been linked to a number of deadly health effects, including heart attacks — has remained about the same over the past five years. Levels of ozone, a gas created when nitrogen oxides released from coal plants react in the presence of sunshine and other heat sources with other pollutants, remains a serious problem in many regions of the country, especially during hot summer months. In 2004, 29 million children ages 14 and under lived in counties with unhealthful ozone levels.

Race is also a factor in exposure to air pollution. A recent study by the Clean Air Task Force, a Boston-based group known for its level-headed research and public policy advocacy, found that about 70 percent of African Americans and Latinos live in counties that violate federal air pollution standards for one or more pollutants. Not surprisingly, the states with the highest per capita mortality rate from power plant pollution — West Virginia, Kentucky, and Tennessee — are all ringed by dirty coal plants.

A deeper understanding of the harmful effects of air pollution also puts the trend toward cleaner air in perspective. In recent years, researchers have learned that particulate matter is damaging not only to our lungs but also to our hearts and perhaps even to our brains. They have learned that long-term exposures to low levels of air pollution can be as dangerous as short-term exposures to high levels of pollution, and that some people — especially the young, the old, and those with medical conditions such as diabetes and heart problems — are far more vulnerable than the general population. We also know that pollution can have unexpected long-term health effects. For example, children who grow up in areas with high levels of ozone pollution have smaller, weaker lungs.

The key debate today is not whether pollution from coal plants kills people. It indisputably does. But so does pollution from cars, trucks, and school buses. We live in a stew of chemicals, some more toxic — and more mysterious — than others. One definition of progress is how well we minimize the risks and maximize the gains. New coal plants being built today are unquestionably cleaner than old coal burners like Hatfield's Ferry, but they still emit large quantities of pollutants into the air and water. And as these new coal plants are built, new questions arise along with them: How much pollution is "safe"? How well do we really understand the health effects of power plant pollution? And perhaps most important, how much risk are we, as a society, willing to take in pursuit of cheap energy?

Big Coal, of course, employs squads of researchers, lobbyists, and friendly politicians who argue that mercury is toxic only if you consume a thermometer full of it and that it's particles from diesel engines, not coal plants, that kill people. Their basic message is this: *We're spending billions on pollution controls, we're responsible citizens, we're obeying all the laws, and now, thanks to all our hard work, coal no longer deserves its reputation as a dirty, nasty fuel.* To paraphrase one uncharacteristically blunt Ohio coal operator, "I'll bet you won't find any dead people near those power plants along the river."

In the fourteenth and fifteenth centuries, the sulfury smell of coal was associated with brimstone, long known as the building block of the underworld and often affiliated with divine wrath. (In Genesis, God destroys Sodom and Gomorrah in a rain of fire and brimstone.) During the Black Death, one Welsh writer wrote of "death coming into our midst like black smoke." In the early days of the industrial revolution, rising cities such as Manchester, England, were literally swallowed up by coal smoke. The life expectancy of a well-to-do member of the Manchester gentry was only thirty-eight years; for a child born in poverty, it was only seventeen years. Many children were deformed by rickets — their growth stunted, their legs and spines curved — due to the almost total lack of sunlight in the smoke-shrouded city.

But during the industrial revolution, smoke also was a sign of progress and sophistication. "Smoke is the incense burning on the altars of industry," one Chicago businessman said in 1892. "It is beautiful to me." A British journalist, visiting the United States in 1912, remarked on a train ride through Ohio, "From a final cat-nap I at last drew up my blind to greet the oncoming day, and was rewarded by one of the finest and most poetical views I have ever seen: a misty, brown river, flanked by a jungle of dark reddish and yellowish chimneys and furnaces that covered it with shifting canopies of white steam and of smoke, varying from the delicatest grays to intense black; a beautiful dim gray sky lightening, and on the ground and low, flat roofs a thin crust of snow: Toledo!"

Cities such as Pittsburgh and Baltimore, where residents burned bituminous coal for home heating and cooking, were much dirtier than cities such as New York, where people burned cleaner anthracite. In an oral history of Atlanta, Dr. Leila Denmark, one of the first women physicians in the city, remembered what it was like to practice medicine in the soot-filled environment:

When I moved to Atlanta [in 1926], by ten o'clock you had a moustache. They all burned soft coal, all the heat they had was soft coal. There was so much smoke that you'd inhale it and your upper lip would be black. And we used lots of ultraviolet light in our office on little children because they got no sunshine . . . You could smell sulfur all over the place. It was really bad. And you'd see an autopsy at Grady Hospital, or any place you'd see the autopsy, if the person was a pretty good age, the lungs were striped. It looked like a zebra! They'd inhaled so much carbon that the lung was black, had black streaks in it.

Many of America's first smoke fighters were women. Tired of soot-fouled laundry and concerned about the lack of sunlight and the wheezing and coughing of their children, they organized into smoke-abatement groups that staged rallies, pressured politicians, and generally put the issue of coal smoke on the public agenda for the first time. Historian John Stilgoe points out that it was not until engineers learned that decreasing smoke saved money, however, that the nation's smokestacks began to clear. Dark smoke signaled incomplete combustion; more pollution meant they were wasting fuel. The motive was profit, not public health.

By the early 1940s, coal was on the decline, replaced by cheaper, cleaner, more flexible fuels. Locomotives switched to diesel, and residents in cities such as Pittsburgh began burning natural gas to heat their homes. But industrial pollution, especially from steel and coke plants, was much slower to improve. (Coke, a fuel that is commonly used in steel production, is produced by baking coal in superhot ovens to drive off impurities and concentrate the carbon.)

The real awakening to the dangers of air pollution didn't come until October 26, 1948. On that morning, a blanket of cold air settled over the industrial town of Donora, Pennsylvania. The chilly air created an atmospheric inversion, trapping gases from Donora's mills, furnaces, and stoves in the valley and creating a blinding fog of coal, coke, and metal fumes. Within days, half the town fell ill. The local funeral home ran out of caskets. The basement in the community center, where Girl Scouts and Boy Scouts usually met, became a temporary morgue. Twenty people died during the fog, and thousands fell ill. (Fifty more deaths were later attributed to the smog.) In 1952, a similar temperature inversion occurred in London. The sulfury fog was so bad that trains and buses crashed and a ferry rammed another vessel in the harbor. Visibility in the city was reduced to eleven inches. Twelve thousand people died.

These "killer fogs" proved what people had suspected for years: high levels of air pollution are deadly. But what about lower levels of pollution? Most public health officials at the time went along with the toxicological adage that the dose makes the poison. Clearly, people weren't dropping dead every time someone lit a coal fire. Exactly where officials drew the line between annoyance and public health emergency was unclear, but it seemed obvious that lower levels of coal smoke were relatively harmless — a few coughs, maybe some sooty phlegm, but nothing a healthy body couldn't handle.

Nevertheless, the killer fogs helped inspire a new awareness of the dangers of air pollution, both from coal plants and vehicles. A similar awakening in California, where tailpipe pollution sometimes reduced visibility in the San Fernando Valley to just a few feet, inspired some of the nation's first air pollution laws. In 1947, California governor Earl Warren signed the Air Pollution Control Act, authorizing the creation of an air pollution control district in every county of the state, the first such law in the nation. In 1969, California became the first state to pass air quality standards for key pollutants, including particles, sulfur dioxide, and nitrogen dioxide.

By that time, the broad environmental degradation of rampant industrialization could no longer be ignored. On New Year's Day in 1970, Republican president Richard Nixon declared the coming decade as a time when "America pays its debt to the past by reclaiming the purity of its air, its waters and our living environment. It is literally now or never." Later that year, after 20 million young Americans massed in parks, dumps, and city streets, waving banners on behalf of our lonely planet to celebrate the first Earth Day, Nixon created a new agency known as the Environmental Protection Agency and signed landmark amendments to the Clean Air Act, both of which signaled the end of the coal industry's free ride. Nixon was no tree-hugger ("You better watch out for those crazy enviros, Bill," he warned EPA head Bill Ruckelshaus in 1972. "They're a bunch of commie pinko queers!"), but he was a shrewd politician, and he understood very clearly that any political system that fails to protect the health and safety of its citizens will not survive long.

In the aftermath of the killer fogs of the 1940s and 1950s, air pollution researchers made great progress identifying the particular chemicals that made smog — a contraction of "smoke" and "fog" — so deadly. Chief among these chemicals was sulfur dioxide, an invisible gas that is created by the burning of fossil fuels. By the 1960s, it was known that even low levels of sulfur dioxide were an irritant to the lungs, and that high levels of the gas were responsible for many of the deaths in London in 1952. Not surprisingly, sulfur dioxide was one of six pollutants targeted for reductions under the 1970 amendments to the Clean Air Act (the other five were nitrogen dioxide, particulate matter, ozone, carbon monoxide, and lead).

By the late 1970s, however, researchers were beginning to understand that the link between air pollution and mortality was more complicated than it had first appeared. Ironically, one of the most important breakthroughs came not from a medical researcher, but from a young economist and energy analyst named Joel Schwartz.

Schwartz had wanted to be a physicist, or perhaps a mathematician. But after graduation from Brandeis University in the late 1970s, he failed to find a job in physics and ended up working as an energy analyst at the EPA. At the time, the Reagan administration wanted to halt the phaseout of lead in gasoline. One of Schwartz's supervisors asked him to produce a study showing how much money the industry would save if the phaseout was halted. Schwartz did it, and then, entirely on his own, he worked up a study showing that the $100 million or so savings to the industry would be more than offset by health costs that could total more than $1 billion annually. In 1982, on the strength of Schwartz's study, the administration reversed its position on the lead phaseout and actually tightened standards.

Schwartz then became interested in particulate air pollution. He had a hunch that these tiny gobs of carbon, metals, and acids created by burning fossil fuels might be more important to respiratory health than many people thought. Because the impact of air pollution on any one person is likely to be complex and difficult to identify, epidemiologists often studied large populations to see whether mortality corresponded with changes in pollution levels. Schwartz's secret weapon was his mastery of complex time-series analysis — an examination of how one variable changes in relation to another over time — which was most often used by econometricians to track inflation or the national debt. In his analysis, Schwartz was able to control for a variety of confounding factors, including smoking and high blood pressure, allowing him to track more precisely the relationship between individual pollutants in the air — sulfur dioxide, nitrogen dioxide, and particulate matter — and death.

Schwartz's method grew out of the understanding that there is a big difference between the risk of health effects from pollution to any one individual and the risk to an entire community or nation. For instance, researchers now estimate that every ten microgram per cubic meter increase of particle pollution increases the death rate in the exposed population by .05 percent. For any one person, that's a minuscule risk, about on par with getting hit by lightning while you're pushing a grocery cart across the parking lot on a Sunday afternoon. But spread that .05 percent over millions of people, and you end up with thousands of dead bodies.

Schwartz decided to use his method to reassess data from the London killer fogs of the 1950s. Those studies had correlated premature deaths and illness with air pollution in London, but had concluded that sulfur dioxide was the killer. Schwartz's analysis proved that conclusion wrong. He found that particulate matter was more deadly than sulfur dioxide (although sulfur dioxide, which reacts with water in the atmosphere to create sulfuric acid, contributes to the formation of dangerous particulates). Again, Schwartz's work led to changes in the law. In 1987, the EPA established standards for particle pollution in the United States. For the first time, the agency began tracking and regulating the level of particles that were smaller than ten microns in diameter (PM_{10}) — about one seventh the diameter of a human hair. Scientists believed that these small particles were responsible for most of the deadly health effects of air pollution, and the law was lauded as an important step forward in protecting public health.

But something continued to bother Schwartz about his London research. No matter how he analyzed the relationship between hospital admissions, deaths, and levels of particle pollution in the air, he could find no threshold of particle levels at which people no longer got sick. He began to worry that the EPA's PM_{10} standard was too coarse; smaller particles might be equally dangerous, and they were both invisible and unregulated.

Schwartz's concerns led him to Steubenville, Ohio, a town that is notoriously polluted by steel mills and other industrial sources of air pollution. He evaluated death certificates in Steubenville and their relationship to levels of particulate matter over a ten-year period. He discovered that deaths from pneumonia, lung disease, and heart attacks rose along with increases in particle pollution, even on days when particle levels were "safe" as defined by the 1987 standards.

Soon Schwartz collaborated with Douglas Dockery, an epidemiologist at Harvard's School of Public Health who shared Schwartz's interest in particles, on a study of the health effects of particle pollution in Philadelphia. Their results mimicked those from Steubenville. Again, there seemed to be no threshold below which particle pollution ceased causing illness and death.

While Dockery and Schwartz were looking at eastern cities, another researcher, C. Arden Pope at Brigham Young University, was making similar observations in Utah. In the mid-1980s, Pope recognized "a unique, natural experiment" when the Geneva Steel mill in the Utah Valley shut down during a labor dispute, allowing Pope to explore the health effects of pollution from the plant. Geneva Steel was the focus of tremendous anger in the region. Residents blamed the black plumes from its smokestacks for making their children sick. Pope discovered that children's hospital admissions for respiratory disease essentially doubled during periods when the mill was operating versus when it was not.

Dockery and Pope then began collecting data on what was to become known as the Harvard Six Cities Study, which included Steubenville, the most polluted of six selected cities, and Portage, Wisconsin, the least polluted. The scientists interviewed eight thousand people about their smoking habits, diet, weight, and other risk factors, taking "lifestyle" differences into account, and tracked deaths among them for fourteen to sixteen years. Among other things, the study showed that the annual death rate was 26 percent higher in Steubenville than in Portage. It also found that of all air pollutants, fine particles had the highest correlation with death rates and that the health effects of fine-particle pollution accrued over the years. Later, another exhaustive study by Pope and Dockery reaffirmed these findings and, for the first time, linked particulate matter with fatal heart attacks.

In 1994, Schwartz was awarded a MacArthur Foundation "genius grant," the first federal employee to be so honored. Shortly thereafter, he abandoned government service for Harvard's School of Public Health. "It was getting harder and harder to do this particle research [at the EPA]," Schwartz explained. "The Republicans were running Congress, and there was a lot of harassment." As for his motivation, Schwartz said, "When I look at the statistics, I see more people dying of particle air pollution than are dying of AIDS, and I need to call people's attention to that."

In 1996, when EPA director Carol Browner finally proposed regulations for fine-particle emissions, the fossil fuel industry went on the attack. The crusade was headed by a group called the Air Quality Standards Coalition, which received large contributions from the National Mining Association ($312,000) and American Electric Power (AEP), the big midwestern utility ($110,000). The group was led by C. Boyden Gray, former counsel to President George H. W. Bush and now a powerful lobbyist. With the help of a slick Washington, D.C., PR firm, the group organized faux populist rallies around the country and ran ads suggesting that the new rules would result in traumatic lifestyle changes, including an eventual ban on lawn mowers and backyard barbecues. ("Tell the EPA that barbecuing is not a crime!" read one of the group's bumper stickers.) The group contended that the passage of these laws would lead to economic ruin for the country. Industry estimates ranged from $23 billion to $80 billion to comply, while Gray himself put the number at closer to $200 billion. In regard to the proposed rules, Virginia governor George Allen told a meeting of Republican governors, "You look at the map and there are states that will be flat out of business."

The EPA disagreed, estimating that the new standards would cost industry between $6.5 billion and $8.5 billion. At the same time, they would save the country an estimated $51 billion to $112 billion annually in medical costs.

The industry group eventually decided that fighting over numbers was a losing battle. ("If we've learned any lesson, it's that you have to engage the debate on a different basis than costs," AEP's John McManus told the *New York Times*.) Instead, the group decided to try a novel approach: attack the scientists. According to a report published by the Center for Public Integrity, a government watchdog group in Washington, D.C., notes kept by one member of the Air Quality Standards Coalition who attended a meeting on February 6, 1997, made it explicit: "Spent a lot of time on Schwartz. [Boyden] Gray spoke of impeaching his reputation, later changed to discrediting him. This crowd feels Schwartz is 'the enemy,' and can be intellectually embarrassed." Not surprisingly, within days members of Congress were demanding the raw data that Schwartz and Dockery had collected for their research and accusing the EPA of using "secret science."

Despite the group's power and money, they failed to kill the new rules regulating particle pollution. In addition, Schwartz's and Pope's work held up over time and was confirmed by larger studies around the world. Today the electric power industry largely concedes that particulate matter is dangerous but argues that particulate matter from vehicles, which contains more carbon than particulate matter from coal, is the most damaging type. To Pope, it's just more spin. "When you look at the history of air pollution, coal is implicated as early and as clearly as anything," he says. "The bottom line is, both in cities that burn a lot of coal and in cities that make a lot of pollution from mobile sources, you see health effects. What we have learned is, particles from coal are bad, particles from gasoline are bad, and particles from diesel are bad. But exactly how they are bad, and why they are bad — that we don't know yet."

Today researchers are focusing more and more on smaller particles, known as "ultrafines," which are roughly 0.1 micron in diameter, about the size of a single bacterium cell. Ultrafines are important because their size allows them to pass through our lungs' natural defenses and enter directly into our bloodstreams, often carrying whatever metals or acids have glommed onto them right along with them. They are like a FedEx service for all kinds of nasty stuff, the health effects of which researchers are just beginning to understand. "In some ways, it's comparable to smoking," explains Dr. Mark Frampton, a professor of environmental medicine at the University of Rochester in New York and one of the leading experts on particle pollution. "We still don't know the exact mechanism by which cigarette smoke kills people, but we know it does. It's similar with ultrafines. We know they can be deadly; we just don't know exactly how."

Like many other researchers, Frampton is increasingly interested in the way ultrafines are able to induce the systemic inflammation of blood vessels throughout our bodies, perhaps through the stimulation of endothelial cells — a thin layer of cells that surrounds blood vessels and helps to maintain blood pressure — and which causes our blood vessels to constrict. (This constriction can be deadly for people with heart conditions.) Another area of interest, which is even more complex, is the way the inhalation of ultrafines leads to oxidative stress, a kind of internal stampede of molecular fragments that kill healthy cells.

Until recently, most studies of the health effects of particulate matter have focused on the heart and lungs. Now researchers are discovering that ultrafines are able to cross the blood-brain barrier and pass into our brains. Dr. Günter Oberdörster, a colleague of Frampton's at the University of Rochester, recently discovered that within hours after they are inhaled, ultrafines show up not only in our respiratory systems but also in our livers and central nervous systems. Dr. Lilian Calderon-Garciduenas, a neuropathologist at the University of Montana, estimates that about 14 percent of the ultrafines we breathe pass straight into our brains. In one study, Calderon-Garciduenas examined the brains of dogs in Mexico City, one of the most polluted places in North America. She found that the dogs had developed waxy brain plaques similar to those that characterize the onset of Alzheimer's disease. When she autopsied the brains of Mexico City residents (most of whom had died in accidents unrelated to air pollution), she found similar plaques. "Air pollution doesn't cause Alzheimer's," explains Calderon-Garciduenas, "but we may well discover that it accelerates the disease's effects."

When it comes to controlling ultrafines, current EPA regulations for particulate matter may actually be counterproductive. As it turns out, the creation of ultrafines *increases* when larger particles are removed from the mix, primarily because the large particles are like tiny balls of Velcro, causing the ultrafines to adhere to them and drawing them out of the air. Thus, current EPA rules that regulate larger particles could have the paradoxical effect of making the air we breathe more dangerous, not less. "It may turn out that we have been regulating the wrong thing," Dr. Frampton says. "Instead of controlling the mass of the particles that power plants emit, it may be far more effective to regulate the *number* of particles."

If particles are the most deadly air pollutants, mercury may be the most insidious. Unlike conventional pollutants, it is not a gas that dissipates in the air or a microscopic piece of soot that eventually breaks down in the environment. Mercury is an element, one of the ninety-four naturally occurring chemicals that are the building blocks of our world. Mercury does not erode or decay; once it is released into the environment, it stays there. Mercury is also highly toxic. In 1997, Dr. Karen Wetterhahn, a researcher at Dartmouth College, accidentally spilled a single drop of dimethylmercury, a concentrated form of mercury often used in lab research, on her hand. She didn't worry about it — she was wearing latex gloves. She washed the mercury off immediately and didn't think about it again until, five months later, she began bumping into walls and slurring her words. Her doctors were not able to diagnose what was wrong until she told them about the drop of mercury. They did a blood test: mercury poisoning. A few months later, Dr. Wetterhahn was dead.

Coal-fired power plants are the largest emitters of mercury in the United States, releasing about 48 tons, or 96,000 pounds, into the air each year. The mercury is released in three different states: elemental, gaseous, and particle-bound. None of these forms are as potent as the dimethylmercury that killed Dr. Wetterhahn. Nor is the inhalation of mercury released from coal plants known to be a health risk. For most people, the primary exposure to mercury comes from eating fish. One of the unusual characteristics of mercury is its ability to accumulate in the tissues of living creatures. This process often begins when airborne mercury from power plants falls into rivers, lakes, and ponds. Bacteria in the soil transform the mercury into another form, methylmercury, which is then taken up in the plankton that live in the water. The plankton are eaten by little fish, which are in turn eaten by bigger fish, which are then eaten by even bigger fish. All the while, the level of mercury is rising. In some cases, by the time the fish are big enough to be caught and eaten by people, the mercury level can be quite high. Large, long-lived predatory fish such as wild catfish, pike, walleye, large-mouthed bass, and tuna are all common sources of mercury exposure. Canned albacore tuna, the longest-living variety of tuna, has been shown to contain about four times as much mercury as chunk light tuna, which is harvested from younger fish.

Because mercury is a neurotoxin, it is particularly dangerous to the still-forming brains and nervous systems of fetuses and young children. Some studies show that children who are exposed to tiny amounts of mercury in utero have slower reflexes, language deficits, and shortened attention spans. In adults, recent studies have shown an association between increased risk of heart attacks and mercury ingested from eating fish. Some researchers believe that mercury exposure may play a role in the development of Parkinson's disease, multiple sclerosis, Alzheimer's disease, and autism.

Conventional air pollutants are democratic: if pollution levels are high, everyone who breathes is exposed to them. Mercury exposure is different. Exposure comes through eating fish, which means that mercury emitted from a coal plant in Michigan can end up on the dinner plate of a pregnant woman in Nebraska. Also, conventional pollutants are often most dangerous to the elderly and the sick. Mercury is most harmful to the unborn, and its subtle effects may play out over a lifetime. According to a nationwide survey by the Centers for Disease Control and Prevention, one in twelve women of childbearing age already has an unsafe blood level of mercury, and as many as 630,000 babies in the United States could be at risk for health problems. As of 2005, the U.S. Food and Drug Administration and the EPA had issued advisories in forty-four states that warned people to avoid or limit their consumption of certain kinds of fish.

Not surprisingly, Big Coal has waged a campaign in recent years to downplay the effects of mercury from power plants, arguing that only about 1 percent of the total mercury released into the environment comes from power plants in the United States (the majority comes from volcanoes and other natural sources). In addition, because of the way that mercury can be taken up and transported by winds high in the atmosphere, the industry suggests that significant amounts of mercury are deposited in the United States from coal plants in China and India. In other words, Big Coal argues that it's only a very small part of a much bigger problem.

That's true enough, but it's not an accurate description of the situation. According to the EPA, about 30 percent of the forty-eight tons of mercury that American coal plants emit every year falls to the ground in the United States. That's 32,000 pounds of a potent neurotoxin falling into America's land and water. In certain areas, up to 80 percent of the deposition comes from local sources. And as more and more mercury is released into the environment, it keeps building up. One way to think of coal plants is as giant mercury-excavating machines, taking tons of mercury and other heavy metals that had been safely sequestered underground and recycling them into the air and water. David Krabbenhoft, the leader of the U.S. Geological Survey's National Mercury Project, estimates that the deposition rate of mercury today is three to five times greater than in preindustrial times. As one Pittsburgh doctor who treats kids with high levels of mercury explained to me, "We're soiling our own nest, and we don't have the faintest idea what the long-term consequences will be."

The dangers of mercury were already well-known almost two thousand years ago when the Greek historian Plutarch rebuked a mine owner for using slaves who were not also criminals in his mercury mines. Despite its toxicity, mercury — the only metal that is liquid at room temperature — appealed to early scientists, who were drawn to it for its purity and shape-shifting qualities. In the Middle Ages, alchemists believed mercury was the base metal that could be spun into gold. Isaac Newton, who was fascinated by alchemy and toyed with mercury in his lab, paid for his fascination by lapsing into madness at the end of his life. In the nineteenth century, mercury salts were used to make felt for hats; hatters who breathed mercury fumes all day often acquired a condition known as "hatter's shakes," which left them mumbling and quivering (hence the phrase "mad as a hatter").

The real danger of mercury, however, is how it moves through the environment and up the food chain. Scientists' understanding of this process is relatively new and grew out of a tragedy that occurred in Minamata, Japan, in the early 1950s. Residents of this small fishing town noticed that their cats were acting strangely. "One year, they were sleepy pets; the next, they were hyperactive monsters — screeching, scratching, and jumping around as if possessed," one journalist wrote. Villagers started stumbling as they walked down the street, dropping their chopsticks in the middle of a meal, slurring their words. Then a number of residents began to suffer convulsions or fly into muttering rages. Some slipped into comas and died. In 1956, doctors reported that "an unclarified disease of the central nervous system has broken out." It wasn't until 1959 that the condition was traced back to a surprising source: mercury in fish.

In Minamata, it was no mystery where the mercury had originally come from. Nearly fifty years earlier, the Chisso Corporation, a petrochemical company, had built a factory near the town. In the 1930s, Chisso began using mercury as a catalyst to produce large quantities of acetaldehyde, a chemical used in plastic production. At the end of the day, they simply dumped the mercury into the bay — as much as twenty-seven tons over a period of some thirty years — where it was transformed into methylmercury and began to work its way up the food chain. For years, Chisso denied there was any link between mercury and the sickness and deaths of residents, preferring to pay off victims in an attempt to keep them quiet rather than stop dumping. It wasn't until 1968, when babies were born with gnarled limbs and public outrage began to boil over, that the Japanese government held the company accountable and the dumping ceased. Today nearly 1,500 people are officially estimated to have died due to mercury poisoning in Minamata Bay, with some 15,000 injured or damaged. Unofficial estimates are three or four times higher.

The disaster in Minamata raised urgent questions about the presence of mercury in the environment and its toxicity. Where else might high levels of mercury be found? Who was at risk? As early as 1975, it was clear to Japanese scientists that there was no "safety level" for mercury. "The greater the methylmercury intake, the greater the cell damage. The lower the intake, the less damage to cells," Dr. Masazumi Harada wrote. "On the cellular level, there is no 'threshold point.'"

Despite the obvious dangers to public health, laws that restricted mercury releases were slow to take shape. In 1990, under President George H. W. Bush, Congress amended the Clean Air Act to strengthen regulation of air toxins including heavy metals such as lead, mercury, arsenic, and cadmium. Under the new law, sources of these air toxins that were found to pose a threat to public health must be controlled by what the Clean Air Act calls maximum achievable control technology, or MACT. But it was another seven years before an EPA report to Congress identified mercury as the health hazard of greatest concern from power plants. Another three years passed before an exhaustive study by the National Academy of Sciences, the most respected scientific body in the country, essentially agreed with the EPA's findings, determining that children of women who eat fish and shellfish regularly during pregnancy are at risk for permanent neurological damage. (Today EPA scientists estimate that one in six women of childbearing age has enough mercury in her blood to put her child at risk should she become pregnant.) Finally, in the waning days of the Clinton administration, the EPA issued a regulatory determination stating unequivocally that power plants are a major source of hazardous air pollutants and that it was appropriate and necessary to regulate mercury emissions under the Clean Air Act. In announcing its decision, the agency noted that "mercury emissions from power plants pose significant hazards to public health and must be reduced." The deadline for proposing new mercury regulations was set for December 15, 2003.

Initially, Big Coal kept quiet about the upcoming ruling. In the fall of 2000, Quin Shea, a lobbyist for the Edison Electric Institute, the voice of the electric power industry, had tried to nudge Big Coal into seeing the political virtues of accepting the new regulations. "If we cannot convince the public that electric companies are serious about addressing environmental issues," Shea wrote in an article for Edison Electric Institute members about the new mercury rules, "we stand to lose our credibility as an industry and to open the door to further attacks against coal-based generation."

Then the George W. Bush administration took over, and within a few months, lobbyists such as Shea began singing a different tune. At a private meeting of western coal producers at a resort in Santa Fe, New Mexico, Shea argued that the coming mercury regulations would increase permitting time for new coal plants by a year or more, potentially stalling the boom in new coal-fired power plants they were all counting on. "Mercury is a killer," Shea told the crowd, referring not to children, but to hopes for a new generation of coal plants in America.

But now that the EPA had already issued its ruling, it was not clear that anything could be done. The only way to get around a crackdown on mercury from coal plants would be to delist the plants as sources of hazardous air pollutants. To do that, the EPA would have to argue that the emissions of hazardous pollutants *from any one source* were low enough that the public health was protected with an ample margin of safety. Given what the nation's best scientists had already determined about the dangers of mercury, and given that a single coal plant can pump one thousand pounds of the toxin into the air in a single year, that would be an extremely tough sell.

Initially, attempts to stall the implementation of new mercury regulations were caught up in the larger politics of clean air. In February 2002, President Bush announced his Clear Skies Initiative, a sweeping proposal to cut emissions of three main pollutants — sulfur dioxide, nitrogen oxides, and mercury — by 70 percent by 2018. According to the administration, Clear Skies would cost the industry about $6.5 billion for new pollution controls but yield $93 billion in health benefits.

Politically, the most compelling aspect of Clear Skies was its embrace of cap-and-trade programs to regulate these pollutants. Cap-and-trade is an innovative emissions trading scheme that was first deployed under the 1990 Clean Air Act amendments to reduce the levels of sulfur dioxide from power plants. To put it simply, it gave each power company a cap, or quota, for a plant's sulfur dioxide emissions, then allowed the company to meet the cap either by installing scrubbers, changing the type of coal the plant burned, or purchasing credits from another company that had reduced its emissions beneath its quotas. The idea was to set up a system that would encourage creativity and efficiency rather than lawsuits and regulatory stalemates. It had worked well for sulfur dioxide, resulting in emissions reductions that were cheaper and occurred faster than anyone had predicted. The EPA projects that the amendments will reduce the death rate from air pollution in 2010 by 23,000 people per year. With Clear Skies, Bush essentially proposed revising and expanding the sulfur dioxide trading program to include nitrogen oxides and mercury.

Despite the initiative's progressive appearance, there were a number of problems that limited Clear Skies' appeal. The biggest issue was that the proposal didn't include any limits on carbon dioxide emissions, which many environmentalists and clean power advocates believed was necessary in any sweeping regulatory reform. Another problem was the loose caps on sulfur dioxide, which would allow higher levels of pollution than mandated under current law. Finally, there was the whole issue of allowing emissions trading of a toxic chemical like mercury. Flexible trading mechanisms worked well for pollutants such as sulfur dioxide, which disperses quickly in the air and has a relatively short lifespan. But a trading system had never been used for a toxic chemical such as mercury before, and many environmentalists and public health advocates believed it was not appropriate to use it now. They feared that a cap-and-trade program for mercury would allow big emitters in effect to buy their way out of making reductions, which would be likely to create local "hot spots" where mercury deposition was high. More important, the cap wasn't very tight. Under Clear Skies, the industry would have until 2010 to cut mercury emissions from forty-eight tons to twenty-six tons, and then until 2018 to cut emissions to fifteen tons. In contrast, if mercury were regulated under the MACT requirements of the Clean Air Act, which mandated the latest control technology at every plant, mercury emissions would fall to about five tons by 2007.

For Christine Todd Whitman, Bush's choice for EPA administrator, mercury was hardly a front-burner issue. Bush's decision to appoint Whitman to head the EPA had surprised many Beltway insiders, who hoped for a lower-profile appointment. Whitman, who had the highest name recognition of anyone in the cabinet except Colin Powell, had been the popular governor of New Jersey for seven years. Pro-choice, fiscally conservative, and socially moderate, she seemed to have a bright political future and was often touted as a possible vice-presidential candidate. During her tenure as governor, Whitman had had a mixed record on environmental issues, but she was not a knee-jerk ally of Big Coal. In fact, she had led the fight to get midwestern power plants to clean up the pollution that drifted into New Jersey and had proposed state legislation aimed at cutting greenhouse gas emissions.

Once she took over at the EPA, however, Whitman quickly found herself marginalized by the power and influence of Vice President Dick Cheney and the former energy executives and lobbyists who dominated the Bush administration (more about Whitman's troubles in a later chapter). While Whitman went through the motions of lobbying Congress for Clear Skies, she rarely mentioned mercury. Bush rarely spoke of it either, and despite his initial enthusiasm for Clear Skies, he never spent a nickel of political capital on it, leading many critics to suggest that the proposal was intended merely as a fig leaf to lend some modesty to the administration's industry-friendly agenda.

When Clear Skies failed to gain momentum, Big Coal turned its full attention to shaping the new mercury rules that would be announced in late 2003. A key part of this strategy was downplaying the dangers of mercury toxicity. For this, Big Coal's argument rested almost entirely on the conclusions of one study. Researchers from the University of Rochester have spent years testing the neurological development of people in the Seychelles, a group of islands in the Indian Ocean. These islanders' diet is made up of high amounts of mercury-tainted seafood. So far, the researchers have found no effect on the neurological development of children from prenatal exposure to low levels of mercury. Why is unclear; it could have to do with genetic factors or the rest of their diet. Although the National Academy of Sciences did not find any serious flaws in the study itself, it's worth noting that parts of the study were funded by a $486,000 grant from the Electric Power Research Institute, which has a long history of favoring research that supports the industry's point of view.

The conclusions of the Seychelles study are contradicted by virtually the entire body of mercury toxicity research. As Dr. Deborah Rice, a former EPA toxicologist, testified before Congress in 2003, "At least eight studies have found an association between methylmercury levels and impaired neuropsychological performance in the child. The Seychelles Islands study is anomalous in not finding associations between methylmercury exposure and adverse effects." The most important of these other studies was conducted in the Faeroe Islands, near Denmark, where mercury-tainted pilot whales are a staple of the islanders' diet. Philippe Grandjean, an epidemiologist at Harvard's School of Public Health, tested a large group of Faeroese children at the ages of seven and fourteen and found that IQ dropped 1.5 points for every doubling in prenatal exposure to mercury. "We have learned there is a response at low levels," Grandjean says. "It's not huge, but it's certainly not negligible."

As the December 2003 deadline for the new mercury rule approached, the noise level from Big Coal supporters began to rise. Senator James Inhofe, the Oklahoma Republican in charge of the powerful Environment and Public Works Committee and best known for his now infamous characterization of global warming as "the greatest hoax ever perpetrated on the American people," presided over a Senate hearing about the new mercury rules. Inhofe went out of his way to praise the Seychelles study, while giving scant play to studies that contradict it. The U.S. Chamber of Commerce, a longtime ally of Big Coal, tried to blame tree-huggers: "The push to regulate mercury emissions from power plants is an attempt by extreme environmental groups to hinder economic growth and force jobs overseas."

During congressional testimony over the new mercury rules, Big Coal used similar rhetoric, arguing that the technology to remove mercury wasn't ready for prime time yet and that cutting mercury emissions would lead to coal plants closing down and the loss of tens of thousands of coal mining jobs. One coal industry expert suggested that to achieve a 90 percent reduction of mercury from coal plants would cost the industry $19 billion and result in, at best, minimal public health gains. As usual, this gloom and doom was widely disputed, not just by environmental groups but also by state regulators and the manufacturers of mercury control devices, who said that without laws capping mercury emissions, power companies had no incentive to develop better control technologies. Innovation follows regulation, they argued, not the other way around. They pointed out that activated carbon injection, a technology that had been tested on a number of coal plants and had been used for years on municipal waste incinerators around the country, had achieved mercury removal rates of more than 90 percent. This technology was improving and prices were falling rapidly. According to one report by state air pollution regulators, installing scrubbers that would remove 90 percent of the mercury from power plants' emissions would add just 15 to 60 cents a month to the typical residential electric bill.

Truth be told, not everyone in the coal business was against tough mercury emissions standards. Many eastern coal companies saw these standards as a way to gain a competitive advantage over their rivals in the West. Because of variations in the chemical structure of coal from different regions, it's often easier to remove mercury from eastern coal than from western coal. Also, because eastern coal usually has a higher heat rate, there is less mercury per unit of energy in eastern coal than in western. In fact, the uproar against tough mercury regulations was largely driven by a small number of western coal companies, as well as by the railroads that haul western coal and the companies that burn it.

A few weeks before the December deadline, rumors leaked out that the EPA was about to make an almost unthinkably bold move: it would propose removing coal plants from the list of sources of hazardous air pollutants, allowing the agency to dump the tough MACT regulations and replace them with a softer, more flexible cap-and-trade proposal that the industry preferred. In effect, having failed to rally support for Clear Skies in Congress, the administration decided to push it through regulatory changes. (The administration did something similar with sulfur dioxide and nitrogen oxide reductions, proposing the Clean Air Interstate Rule, or CAIR, which was essentially a repackaging of Clear Skies.) And if that weren't enough, the caps in the new mercury proposal were even looser than what had been proposed in Clear Skies: instead of cutting mercury emissions to twenty-six tons by 2010, the new proposal capped them at thirty-four tons, with a final reduction to twenty-six tons by 2018.

Why thirty-four tons instead of twenty-six tons? According to one EPA researcher who worked on the proposal, it was simply the number that popped up on the agency's computer models when they calculated the mercury reductions that would be gained as cobenefits from new pollution controls for sulfur dioxide and nitrogen oxides required under CAIR. The number, to be blunt, had nothing to do with public health and everything to do with making life easy for Big Coal. The way the rule was structured, it would be a good twenty years before the industry had to get serious about cutting mercury. A more important question was this: how did the EPA justify the decision to delist coal plants as a source of hazardous air pollutants? Did the agency present compelling new science that showed that mercury emissions were not as dangerous as previously believed? No. It simply interpreted the data in previous studies and overturned the conclusions of the nation's best scientists. The agency asserted that the "EPA, in its expert judgment, concludes that utility [mercury] emissions do not pose hazards to public health."

The credibility of this "expert judgment" was destroyed a few weeks after the proposal was released when the *Washington Post* reported that at least a dozen passages in the EPA's proposal were lifted, sometimes verbatim, from memos prepared by West Associates, an industry organization representing western coal burners, and Latham & Watkins, a powerful Washington law firm that often represents corporations on environmental issues and where EPA air policy chief Jeffrey Holmstead once worked. Holmstead tried to dismiss the matter as nothing more than an interagency mix-up: "That's typically not the way we do things, borrowing language from other people," Holmstead told the *Post*.

In the following months, the Government Accountability Office, the congressional watchdog agency, blasted the EPA for in effect cooking its books, including rigging the comparison between the original MACT proposal and cap-and-trade, as well as failing to adequately assess the health benefits of cutting mercury emissions. But even without an accurate analysis, the EPA admitted that the tougher MACT standards would lead to net economic benefits — that is, benefits *minus* costs — of $13 billion. The benefits of the cap-and-trade proposal would undoubtedly have been lower, but the agency never bothered to calculate them in a comparable way.

Why would they? The point was never to weigh the evidence and make the best decision. The point was to figure out what they could get away with and ram it through. At times, the tactics used to get this done bordered on intimidation. After the Georgia Environmental Protection Division sent a letter to the EPA opposing the mercury rollback, the agency received a personal visit from none-too-pleased representatives from the Electric Power Research Institute. "They treated us to a three-hour lecture on mercury," Dr. Randy Manning, the top toxicologist with the Georgia Environmental Protection Division, told me. "The gist of it was, *You're morons; you don't know anything.*"

The EPA's mercury rule was finalized in March 2005. Within weeks, fourteen states filed suit to have the new rule overturned, charging that the cap-and-trade scheme was unlawful under the strict requirements for the regulation of hazardous air pollutants in the Clean Air Act. The courts may eventually force the EPA to dump the new rule and go back to developing a regulatory standard for mercury that does not flout the law. But by that time, another five years will have passed and a new generation of coal plants will be under construction.

For Big Coal, the obvious question is this: was the fight over mercury worth it? Yes, the industry won a decade or more of delay before it has to install mercury controls on power plants. But in return, it received mountains of bad press, provoked passionate anti-coal outbursts at public meetings around the country, triggered more than a dozen lawsuits, and inspired nearly half a million protest letters to the EPA. One might easily argue that Quin Shea had been correct back in 2000: by engaging in a political street fight over mercury, Big Coal made a mockery of its own PR rhetoric about being "increasingly clean" and undermined whatever credibility it had as an industry willing to face up to tough problems.

But that's not quite the whole picture. For Big Coal, the fight over mercury was also about staking out a favorable position in the next battle over clean air. By getting coal plants delisted as a source of hazardous air pollutants, it also cut off any future effort to limit the release of other toxic heavy metals that coal plants emit, such as arsenic, chromium, manganese, beryllium, and lead. For Big Coal, this is no small achievement. Between 1975 and 2001, the annual releases of toxic metals from coal plants are estimated to have nearly doubled, from about 350 tons to 700 tons. Toxic air emissions from power plants dwarf those from all other industries in America, accounting for more than 40 percent of all air toxins reported to the EPA. And yet, unless the delisting of coal plants as sources of hazardous air pollutants is overturned by the courts, coal plants are no longer legally considered a source of these toxins and cannot be regulated.

The health effects of these heavy metals are poorly understood, in part because their interaction with the human body is exceedingly subtle and complex. But what is known is not encouraging. Human skeletal lead burdens today are five hundred times those of ancient Peruvians, who did not use lead for smelting. Contemporary bone levels of cadmium, which, like lead, is virtually absent from humans at birth, are about fifty times higher than those found in Pecos Indians of the American Southwest circa A.D. 1400. Some studies have shown that violent and nonviolent incarcerated male criminals differ significantly in cadmium and lead levels. Other studies have shown that high cadmium and magnesium levels characterize disruptive U.S. Navy recruits. A number of researchers believe that the rise in developmental disabilities such as dyslexia, attention deficit hyperactivity disorder (ADHD), mental retardation, and autism, which now affect between 3 and 8 percent of all children in America, may be related to exposure to heavy metals. In the coming years, researchers may well determine that toxic metal emissions from burning coal have little to do with the rise in these developmental problems. But as Jesse Ausubel, the director of the Program for the Human Environment at Rockefeller University, wrote, "I worry that the Industrial Revolution, which has spared us from stoop labor and the hazards of the fields and brought us comfort, convenience, and mobility barely imaginable two centuries ago, has poisoned many among us."

Chapter 7

"A Citizen Wherever We Serve"

WHEN I VISITED the Southern Company's Network Operations Center in Birmingham, Alabama, I was not thinking about how Big Coal wields its political power. I was simply trying to understand how electrons move from one place to another. Southern's operations center, which is located in a secure, bombproof location under a lovely art deco building downtown, seemed as good a place to start as any. It is the electronic hub of the Southeast, where the power flowing over transmission lines in Georgia, Alabama, and parts of Mississippi and Florida are monitored and controlled. During my visit, I spent a few hours talking with Mike Miller, whose official job title is "Manager, Operations Planning, Bulk Power Operations" for the Southern Company. In fact, he's the guru of the operations center, a big, broad guy with white hair and a friendly manner. Miller can tell you about how power lines sag when they are overheated and about the headaches caused by a buildup of vulture droppings on the power poles. (If the accumulation gets big enough, it causes electrical shorts.) He can make the idea of moving billions of electrons over the swamps and highways of the Southeast seem only slightly more difficult than planning a fancy dinner party.

At a certain point in our conversation, Miller and I began to talk about the blackout in the Northeast in the summer of 2003. Miller is justifiably proud of the fact that the Southeast has not had a major blackout in recent memory, and he spends a good part of each day trying to make sure it doesn't happen in the near future. "Right now, in the aftermath of 9/11, people want normalcy," Miller told me. "Keeping the lights on is more important than ever. I think politicians understand this, too. They understand that a blackout, even a small one, would add to the perception that our lives are in danger, that we are losing control. Control is very important right now. I think George Bush knows this as well as anyone." Then Miller added matter-of-factly, "Electric companies have the power to swing an election."

Miller did not mean this as a threat. It was simply a blunt and rather astonishing statement of the political muscle of the electric power industry. No other industry in America has anywhere near this kind of leverage over our public and private lives. Even OPEC, the Middle Eastern oil cartel, pales in comparison. If OPEC wants to throw its weight around, it can cause economic chaos. But the power companies can shut down our lives with the flip of a switch. Not that they ever would. But the fact that the subject even comes up in conversation is telling.

If there is one company that understands this political dynamic better than any other, it's the Southern Company. Southern is a direct descendant of Commonwealth & Southern, which, like Insull's Middle West Utilities Company, was broken up after the passage of the Public Utility Holding Company Act of 1935. In some ways, Southern hasn't changed much since the glory days of Insull. The company is still burning coal, still touting cheap power, and still positioning itself as a servant of the community while running headlong over regulators. Southern's holdings include a retail gas business, a nuclear plant subsidiary, a wireless company, a fiber-optic business, a wholesale power generation business, and, at the center of it all, five electric utilities — Georgia Power, Alabama Power, Mississippi Power, Gulf Power, and Savannah Electric. Together, these utilities serve more than 4 million customers in four states. Southern's seventy-nine generating stations produce about 39,000 megawatts of electricity, about 70 percent of which comes from coal. It is literally an empire built on black rocks. Even the Georgia Power building, one of Atlanta's most visible corporate landmarks, with its odd trapezoidal shape and sleek all-black glass exterior, resembles nothing so much as a giant hunk of anthracite pushing up out of the pavement in downtown Atlanta.

As many Georgians will tell you, there is much to love about Southern. The company consistently wins top marks in customer satisfaction from research firms such as J. D. Power and Associates. And while many power companies were devastated by the post-Enron meltdown in the electricity sector, Southern remains a financial rock. The company has not missed paying a quarterly dividend since 1948. In 2004, when the readers of *Fortune* voted for the electric and gas utility they most admired in America, the winner was no surprise: the Southern Company. The company's CEO, David Ratcliffe, is a folksy, white-haired man who has spent nearly his entire working life at Southern and exudes a quiet, confident charm that is the antithesis of famous energy industry hucksters such as Enron's Jeff Skilling. When Ratcliffe takes the stage at the company's annual shareholders meeting and says, in a slow, earthy southern drawl, "We're pleased to tell you that your company is in very good shape," it's hard not to believe him.

As a political force, the company is equally formidable. A lobbyist at a competing power company, long experienced at seeing Southern get its way in Atlanta and Washington, D.C., calls the Southern Company lobbyists "kneecap breakers." Southern's muscle comes, in large part, from money. Between 2001 and 2004, the company spent more than $25 million on federal lobbying fees, far more than any other electric power company in the nation, and more than much larger corporations such as Ford, Chevron, Pfizer, and Monsanto. Southern's political campaign contributions are equally outsize. According to the Center for Responsive Politics, which tracks campaign donations, Southern and its affiliates gave $4.4 million in campaign contributions to federal parties and candidates between 2000 and 2004 — again, far more than any other electric power company and more than well-known corporate goliaths such as ExxonMobil, General Motors, and Halliburton. The *Washington Post* reported that during the 2000 campaign, at least five Bush Pioneers (supporters who brought in at least $100,000 in contributions) were Southern Company executives or lobbyists: Southern executive vice president Dwight Evans; Roger Windham Wallace of

THE SOUTHERN COMPANY'S POLITICAL MUSCLE

Federal Lobbying Expenses 2001–2004

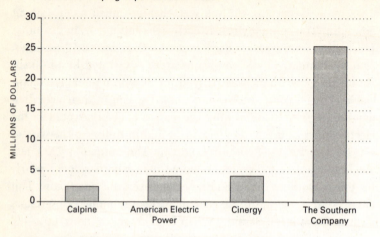

MAJOR U.S. ELECTRIC POWER COMPANIES

Source: Center for Public Integrity LobbyWatch Database

the lobbying firm Public Strategies; Rob Leebern of the firm Trout-man Sanders; Lanny Griffith of the firm Barbour Griffith and Rogers; and Ray Cole of the firm Van Scoyoc Associates. For 2004, this group upped the ante and became Bush Rangers, bringing in at least $200,000 in contributions.

Not surprisingly, Southern's success has made the company the bête noire of environmentalists, many of whom see it as — this is no simple metaphor — getting away with murder. National environmental groups such as U.S. PIRG, the Sierra Club, and the Natural Resources Defense Council have funded report after report with titles such as "Abuse of Power" and "Southern Company: A Giant Among Polluters." These reports list the tons of sulfur dioxide, nitrogen oxides, mercury, and carbon dioxide that the company's power plants spew into the air each year, as well as columns of numbers tallying up Southern's political donations. Unfortunately for these environmental organizations, most of the readers of these reports — as well as the members of these groups — live in the Northeast or the West, where the idea that a company profits by polluting can still muster up some outrage. In the pro-sprawl, pro–gonzo development South, there has been decidedly less outcry.

More surprising is how the company is viewed within the power industry itself. There is, of course, widespread envy of Southern's steady profits, solid management, and excellent service record. But at the same time, it is seen as a prehistoric beast — it is "the *T. rex* of the power industry," says Joe Tondu, an independent power plant builder in Texas — that stomps around the South with virtually no oversight, allowing it to get away with things that no other power company in America would dare attempt. "I think what drives some people in the industry nuts," one consultant told me, "is that they are the antithesis of a capitalist enterprise. They thrive on subsidies, political favoritism, regulatory loopholes, and public ignorance." Another power executive who spent years working with Southern on a failed wind project describes the company with a military metaphor: "They have a trench warfare mentality. Everyone is out to get them. Everyone outside their fiefdom is their enemy."

Southern is brilliant at disarming its critics with good deeds. The Georgia Power Foundation is the third-largest corporate foundation in Georgia, giving some $5 million a year to support cancer research, college scholarships, arts events, cultural diversity programs, and dozens of other causes (any cause, that is, that doesn't have anything to do with asthma, global warming, campaign finance reform, or entrepreneurialism in the electric power industry). Southern also funds educational programming, such as *The Natural South*, a soft-focus TV show about dinosaur bones and salamanders, and supports hunter-friendly "wildlife" programs, such as the restoration of quail habitat on Southern Company land. The company encourages employees to recycle their trash and drive fuel-efficient cars. The first time I visited Georgia Power's corporate headquarters, there were so many flyers and posters urging employees to carpool and bragging about how many pounds of trash the do-gooders at the company had recycled that I thought I'd walked into a Greenpeace office by mistake.

It's easy to see why Southern inspires such admiration — and such distrust. It's a ruthless capitalist with a happy face, a company that's brass knuckles on the inside and a teddy bear on the outside, an organization that has been so brilliant at equating the private good of shareholders with the common good of society that questioning this relationship is a little like suggesting that old Uncle Fred, the beloved relative who bounces the kids on his knee at Thanksgiving, is in fact a pedophile. Even if it's true, who wants to believe it?

The best place to get an up-close look at Southern's political muscle is not in Washington, D.C., but under the gold dome of the Georgia State Capitol in Atlanta. The Georgia legislative session is forty days of big hats, big bellies, and big cigars. During my visit in 2004, raucous debates were held about gay marriage and teaching evolution in schools, as well as less pressing matters such as passing a budget and an ethics bill. As for Southern, its presence wasn't hard to see. Three Georgia Power lobbyists often hung out on the third floor near the entrance to the house and senate chambers, beneath a large portrait of Robert E. Lee. They were always exceedingly well dressed and friendly, happy to exchange hellos at the urinal and chat about the weather, but unwilling to say much more when asked about their work by a nosy journalist like myself.

During the session, I became interested in an obscure bit of legislation that Representative Mike Snow, a Democrat from northern Georgia, was trying to pass. The bill would have prevented utilities from having contact with anyone on the Georgia Public Service Commission (PSC), the state agency charged with overseeing public utilities, while the PSC was working on a case that involved the company. In other words, no hectoring phone calls, no luxurious lunches, no friendly visits to check up on regulators. It was a simple, straightforward reform, prompted by accounts of corporate lobbyists peering over the shoulders of PSC staff and influencing every aspect of the regulatory process. Although Snow's bill was hardly more than a blip on the legislative radar, it seemed like an interesting one to watch, if only because it was such a no-brainer. Unless you worked for the power company, how could you be in favor of allowing the company to meddle with regulators?

Since the days when Insull first established the "natural monopoly" of electric utilities, well-meaning public service commissioners in many states have tried, with little success, to exercise some control over power companies. This is not because their fellow public service commissioners are necessarily corrupt (although they sometimes are), but because they rarely challenge power industry orthodoxy. "Historically, most regulatory officials have shared the primary belief of the power empire that the nation's prosperity rises or falls on the availability of abundant supplies of cheap electricity," one electric power industry critic writes. "The power company views their relationship with the PSC as a 'pact' — the basic agreement: power companies supply a steady supply of electricity and the commissioners make sure they get rates that make them highly profitable and able to borrow increasing amounts of funds from Wall Street for ever-larger construction projects."

For Georgia Power, striking up a friendship with PSC commissioners can pay off big. Consider the 2001 Georgia Power rate case. Every five years, the PSC goes through a long, painful process of deciding what rates electric power customers in the state can be charged. Rate cases are notoriously complex, involving thousands of pages of documents and sophisticated forecasts of future power usage and expenses. In 2001, Georgia Power asked for a $103 million one-year increase. The PSC disagreed, arguing that because of unexpectedly large profits by the utility, Georgia Power customers should actually get a rate cut. Exactly how much was a matter of debate. When deciding a rate case, the PSC staff typically works up two proposals: one by the advocate staff, which usually represents the consumer's point of view, and one by the advisory staff, which represents the industry point of view. In 2001, the advocate staff recommended a cut of $400 million for one year; the advisory staff, in contrast, recommended only a $228 million cut. Months of debate and hearings ensued. Thousands of pages of documents were pored over, analyzed, and discussed. Finally, about a half-hour before the commissioners were set to vote on the plan, the chairman of the PSC, Lauren "Bubba" McDonald Jr., a failed gubernatorial candidate and winner of one Atlanta weekly's annual Golden Sleaze Award, dropped a new forty-page proposal on the desk of commissioner Robert Baker. Unbeknownst to Baker, the power company and the four other commissioners had apparently negotiated a back-room deal. It proposed giving $118 million a year back to ratepayers — about half of what the industry-friendly advisory staff had proposed. Even worse, all but $5 million of that was actually a refund that was due to Georgia Power ratepayers anyway. "It was no more a rate cut than an annual refund check from the IRS is a tax cut," reported Margaret Newkirk in the *Atlanta Journal-Constitution*. "The most ridiculous thing about it was that you could still see the letterhead for Georgia Power's law firm, Troutman Sanders, right on the paper," remembers one official who was involved in the deal. "In effect, they wrote their own rate case." A half-hour later, the new proposal passed by a 4–1 vote, with Baker the only commissioner voting against it. Result: Georgia Power pocketed hundreds of millions of dollars that otherwise might have gone to Georgia ratepayers, and almost nobody noticed.

When I visited Representative Snow during the legislative session in his corner office in the state capitol, he was hardly bashful about expressing his views. "I'd say Georgia Power's influence today is about like the railroads' was fifty or a hundred years ago," he told me. "They pretty much do what they want in this state." Snow is an outspoken character in every way. On this particular day, he was wearing a black-and-white-checked blazer, a blue sweater vest, and a red tie, all of which looked like he'd picked them up at a garage sale on his way into the capitol that morning. As he talked, he munched on a piece of fried chicken. And like most Georgia legislators (the job pays only about $16,000 a year), Snow has a second job. "I'm in the shrubbery business," he said unabashedly.

"I'm all for cheap electricity, but I think Georgia Power has gotten a little too big for their britches in the last few years," Snow told me. He mentioned the 2001 rate case and what an outrage it was. "I'm not saying they're evil, because they're not. I'm just saying that somebody has to start asking some hard questions about how they conduct their business."

Snow, who keeps a photo of himself with President George W. Bush on the wall of his office, doesn't pretend that his motivations are entirely high-minded. "I'm a little angry about some of the games the power company played with me during my last election," he told me over lunch. Snow said that prior to the election, the power company went out of its way to work with his opponent to solve problems with his constituents. "Why would the power company guys do that, except to make me look bad?" Snow also noticed that the company was suspiciously slow to restore power in his region after a line went down. Such meddling in an election might seem trivial, but in a rural district like Snow's, which is heavily Republican, it could make a difference, and Snow took it, right or wrong, as the power company actively working to get him booted out of office.

I asked Snow if he thought his bill would pass — at the time, he was struggling to find more than a handful of cosponsors. "My goal during this session," he said, sucking on a chicken bone, "is to mess with them a little bit. They think I'm just a hick. Maybe I can't get this law passed, but I can make them squirm a little. Why should I be afraid? I'm a fifty-six-year-old guy who sells shrubs for a living. I can afford to be honest."

Several weeks later, the Georgia legislative session ended. Snow's bill not only didn't pass into law, but it never even made it out of committee. So much for messing with the power company.

It wasn't so long ago that life was simple for a big coal burner like Southern. The post–World War II boom in the South, as well as the transformation of air conditioning from a luxury to a necessity, created a demand for power that grew exponentially every year. After Insull figured out the business, all a power company like Southern had to do was build new plants and collect the checks.

Then in the early 1970s, with the Arab oil embargo and the general economic malaise that swept the country, the good times ended. For the first time in the history of the electric power industry, demand plummeted. Many power companies had just embarked on a building spree of a new generation of giant coal and nuclear power plants, which now wouldn't be needed for many years. Southern's biggest subsidiary, Georgia Power, was in worse shape than most companies, in part because it was already several billion dollars over budget on a new nuclear plant. When energy demand collapsed, the company tottered on the edge of bankruptcy. As

Georgia Power's CEO later admitted, "It was a very scary time." The company was saved only by a furious selling off of new power plants (including portions of the yet-to-be-completed Plant Scherer) and a dependably pliant PSC, which allowed the company to boost electricity rates to cover their bad investments.

In the 1980s, electricity consumption picked up again, and Southern's fortunes improved. But the company's troubles weren't over. In 1989, federal agents twice raided the offices of Georgia Power executives, seizing cartloads of documents and computers. The company was investigated by a grand jury for tax evasion connected with an accounting scam in which Georgia Power allegedly overcharged customers millions of dollars for spare parts that were never used. The IRS recommended indictment, but the Department of Justice decided not to pursue the case. (Southern denied any wrongdoing, blaming it all on accounting errors.) That same year, Gulf Power, another Southern Company subsidiary, pled guilty to making illegal political payments and was fined $500,000.

For Southern, the future did not look bright. With the Energy Policy Act of 1992, Congress opened the door to wholesale competition in the electric power industry, which, among other things, meant that Southern's days as a regulated "natural monopoly" were coming to an end. Soon, industry analysts believed, the electric power industry would undergo the same torrent of creative destruction that had ripped through the telephone and airline industries. The prospect of such competition struck terror in the hearts of many Southern executives, most of whom had little experience with market forces beyond vying for tee times at the Druid Hills Golf Club. Compared to slick capitalist machines such as Enron, a natural gas pusher that was then rising out of the Texas flatlands, Southern seemed like an empire from another age, the corporate equivalent of the old, dirty coal plants Southern executives loved so dearly.

This point was driven home by a discrimination lawsuit filed by seven black Southern Company employees, charging widespread bias against African Americans in pay and promotion within the company. In the 1990s, hangman's nooses were found in prominent areas in a number of Southern Company power plants, and a

Christmas card with Ku Klux Klan members singing "White Christmas" had shown up in corporate mailboxes. When the seven employees filed the lawsuit, it was widely seen as preliminary to a much broader class-action lawsuit similar to the one that had rocked Coca-Cola, another Atlanta-based company, in 1999 and had eventually cost the company $192 million to settle. The lawsuit against Southern attracted widespread media attention. CBS News, the *New York Times,* and *Dateline NBC* all covered it. Superstar attorney Johnnie Cochran got involved, perhaps prompted by remarks by Southern's then CEO, Bill Dahlberg, a folksy Georgian with a country boy shtick, who claimed in a deposition that he had "no earthly idea that anybody today would consider [a hangman's noose] to be a racial symbol."

Soon more troubles emerged. On November 3, 1999, Attorney General Janet Reno called a press conference to announce that the U.S. Department of Justice was filing lawsuits against eight utility holding companies that operated fifty-one older coal plants in ten states, charging them with violating a section of the Clean Air Act called New Source Review. Ten of the power plants cited in the lawsuit were owned by the Southern Company — five in Alabama, one in Mississippi, and three in Georgia, including Plant Scherer. (The charges were specifically targeted at violations in the last two of the four units to be constructed at Scherer.) The Justice Department lawsuit cited these plants as "significant contributors to some of the most severe environmental problems facing the nation today." Taken together, the companies named in the lawsuits emitted more than 2 million tons of sulfur dioxide every year and 660,000 tons of nitrogen oxides. "When children can't breathe because of pollution from a utility plant hundreds of miles away," Reno said at the press conference, "something must be done."

New Source Review is an artifact of political compromise. Its roots go back to 1972, when the Sierra Club filed suit against the EPA, charging that the national air quality standards established in the Clean Air Act of 1970 were not working as planned. These standards had set limits not on emissions from power plants and other sources, but on the total pollution level in the air. One of the unforeseen consequences of this law was that power plant operators, instead of installing pollution controls, were complying either by running their plants intermittently (shutting them down when the air got too bad) or by building or retrofitting the plants with taller stacks, which pumped the pollution higher into the air, effectively spreading it over a wider area. Worst of all, the Sierra Club contended, instead of installing pollution controls on power plants in polluted areas, the companies simply built new plants in areas where the air was cleaner and there was, in effect, more room to pollute. The Sierra Club won a decision on the case from a U.S. district court judge, which was subsequently upheld by the U.S. Supreme Court. As a result, in 1974 the EPA issued new rules governing "the prevention of serious deterioration of air quality," which essentially required that all new coal plants be equipped with pollution control devices such as scrubbers. When the Clean Air Act was amended by Congress in 1977, these rule changes, with minor modifications, were written into the law and make up what is commonly referred to as New Source Review.

But what about the hundreds of dirty coal plants that are already chugging away? During the negotiations that led up to the 1977 amendments, power companies complained that it would be too expensive and complicated to install pollution controls on existing plants and argued that control requirements should apply only to new plants. It seemed sensible enough: over the coming years, the old plants would be gradually retired and replaced by newer, cleaner, more up-to-date power plants. By 2005 or so (given the thirty-five-year life expectancy of most power plants), the dirty old coal burners would be gone. It was not exactly an overnight solution, but it was one that both the electricity industry and environmentalists could live with, and a deal was struck. However, the compromise also included the requirement that if an older plant made a major modification, as opposed to performing routine maintenance, it would have to come into compliance with New Source Review at the time of making the modification. The thinking was that as long as the plant was shut down, it might as well install modern pollution control equipment at the same time.

But then a funny thing happened. Instead of retiring the old plants, the power companies decided to keep fixing them up. The reasoning was simple: these old power plants, most of which were paid off and all of which did not have to meet modern pollution control standards, were tremendous profit centers for the power companies, and it made good business sense to extend their lives as long as possible. In addition, the growing public awareness of the dangers of power plant pollution in the 1980s and 1990s gave birth to rampant NIMBY-ism: everyone wanted cheap power but nobody wanted a power plant in his or her backyard. This made newer, cleaner plants virtually impossible to build, which had the paradoxical effect of making the dirty old plants even more valuable. It was the reverse of the Silicon Valley notion of progress, in which new technology was constantly driving out the old. In the case of coal plants, the geographic and economic advantages of old technology drove out the new.

Of course, no one in the power industry explained it this way. Instead, they argued that old power plants were still chugging along because New Source Review rules were so complicated and confusing that it was impossible for utilities to determine the difference between "routine maintenance," which wouldn't require an upgrade, and significant "physical change," which would. There was indeed some wiggle room in the way the law was written. If you wanted to be cynical about it, you could argue that this is precisely what the power company lobbyists and lawyers who helped shape the Clean Air Act had in mind. But it was also true that, over the years, the EPA had issued frequent letters and bulletins telling power companies exactly where the agency was drawing the line. And some of the modifications were hardly subtle. At one power plant in Tennessee, for example, the routine maintenance was so expansive that the company had to build a monorail to haul in new parts.

For the power companies, keeping the old plants running was an economic decision. For the public at large, it was often a matter of life and death. Abt Associates, a respected global research and consulting firm that often works with Fortune 500 companies as well as the EPA, used the EPA's own models to estimate that pollution from power plants owned by the eight companies identified in the lawsuits caused 5,900 premature deaths a year. Of those deaths, 1,200 were attributed to the coal plants owned by Southern.

When the Department of Justice filed the lawsuits, several companies feigned surprise. In fact, the EPA had been investigating violations at the coal plants for nearly three years, subpoenaing hundreds of thousands of pages of documents and meeting with utility officials on numerous occasions. "This was the most significant noncompliance pattern the EPA had ever found," Sylvia Lowrance, a top EPA enforcement official, told the *New York Times*. "It was the environmental equivalent of the tobacco litigation." According to Lowrance, "The companies knew what was going on. A lot of them thought they could evade the law."

Of the eight companies named in the lawsuits, none dug in its heels deeper than Southern. After the lawsuits were announced, Southern issued a statement declaring that it was cooperating with the EPA's investigation and had already provided investigators with more than 120,000 pages of documents. But this was mostly PR. Backstage, the company — with help from a bevy of lawyers at its most trusted law firm, Troutman Sanders — fought hard against the lawsuit, filing research and study papers with the EPA during the comment period. "It was clear from the outset that we were never going to get anything out of Southern," recalls Eric Schaeffer, head of the EPA's Office of Regulatory Enforcement at the time. "During meetings, they'd just glare at us. They were not interested in sitting down to negotiate, not even to fake it. Their body language always sent one message: 'Screw you.'"

Like every other company named in the lawsuits, Southern faced a choice: settle the lawsuit and clean up or shut down its old power plants, or fight it out with lawyers and lobbyists. From a strictly bottom-line approach, this was a no-brainer. If Southern spent $5 million or $6 million a year on lobbyists, that would be a drop in the bucket compared to the several billion dollars or so it would cost over the next twenty years to bring its coal plants up to date.

But little if any of that would come out of Southern's pocket. Because all of Southern's coal plants are operated by regulated utilities, the cost of cleaning up would be passed on to Southern's customers. According to one study, if Southern were to outfit all its coal plants with the latest pollution controls, a typical household in the Southern Company service territory would pay an additional $5.22 a month. For many customers, that's a meaningful sum, and there would surely be protests from consumer rights advocates and cost-conscious industrial users. But as Southern executives know better than anyone, there are ways of mitigating the impact of this price increase on low-income families. In addition, the increased cost of power would be more than offset by the hundreds of millions of dollars in health benefits that Southern's customers would gain from breathing cleaner air. In fact, one study by the EPA found that between 1970 and 1990, for every dollar spent on pollution controls, the public saved $42, mainly in reduced health care costs and increased productivity from people who would otherwise have been injured or killed by air pollution. The company could rev up its spin machine to take credit (discreetly, of course) for the hundreds, if not thousands, of lives that would be saved by cleaning up the old coal plants. Instead of being tarred as a corporate scofflaw, Southern could boast about being a progressive leader in the fight against air pollution in the South and remind its customers that the company's slogan — "A Citizen Wherever We Serve" — is not just an empty phrase coined back in the 1920s but a real commitment to values larger than Southern's annual stock dividends.

After all, Southern Company executives had demonstrated that they were quite capable of doing the right thing. Although Southern's lawyers would eventually succeed in getting the discrimination lawsuit filed against the company in 2000 thrown out by a U.S. district court judge, that didn't stop Southern's management from recognizing that it had a problem with racial bias within the company and moving quickly to remedy it. A diversity action council was formed; employees underwent mandatory diversity sensitivity training; people of color were promoted to supervisory positions. Indeed, fighting racism became something of a personal crusade for then Georgia Power CEO David Ratcliffe, who frequently (some would say relentlessly) addressed the issue in internal speeches to employees. If nothing else, the company's rapid response to the discrimination lawsuit was proof of how quickly Southern's executives could change the corporate culture if they wanted to.

But when it came to cleaning up its coal plants, such urgency and frankness were altogether lacking. It was a lot easier to hire a bunch of lobbyists and slug it out backstage.

In many cases, the New Source Review lawsuits were maddeningly technical, turning on complex distinctions between "routine maintenance" and significant "physical change" of the old plants. But according to Bruce Buckheit, head of enforcement at the EPA at the time the lawsuits were filed, the case against Plant Scherer was remarkably clear-cut: "The suit basically alleges that Georgia Power willingly and knowingly entered into contracts with an intent to evade the Clean Air Act."

For the Southern Company, which was in the midst of planning a number of new coal plants at the time, the implications were enormous. The rules issued by the EPA in 1974 required that any company that began construction of a power plant after June 1, 1975, would be forced to undergo a permit review process to ensure that the plant was built with what the EPA called "Best Available Control Technology." In practical terms, a permit review circa 1975 (requirements have changed in subsequent years) meant that a big coal plant would probably have to install scrubbers to remove sulfur dioxide. And that meant, among other things, an additional expense of $100 million or so on each of the plant's units. (Most big coal plants are made up of a number of smaller plants, or units, each with its own boiler, turbine, and generator.) Given Southern's precarious financial predicament and the sudden decline in demand for electric power in the mid-1970s, this was an expense the company could ill afford. Suddenly, June 1, 1975, loomed large on the horizon: if the company could begin construction before that date, it would have a chance of evading the requirement for scrubbers on the new plant, saving the company hundreds of millions of dollars.

On March 25, 1974, Georgia Power made its first public announcement of its plans to build Plant Scherer. The announcement was hurried; the company still didn't own the land it planned to build the plant on and would spend much of the next year in court fighting to acquire more than thirteen thousand acres from several landowners who had no intention of selling their land to the power company. According to an outside audit that was done in 1987, the contracts weren't signed with the suppliers of the generators for units 3 and 4 until May 23, 1975, just the week before the June 1 deadline. But according to the EPA rules, entering into a contract for construction, as long as it was a legally binding contract with a "significant liability" for cancellation, was evidence of beginning construction. Thus, by the narrowest of margins, Georgia Power ducked under the wire and avoided having to add scrubbers to its new power plant.

Or did it? This question was at the heart of the 1999 New Source Review lawsuits. At the center of these cases is a disagreement over what constitutes "commencement" of construction and how long that construction can be delayed, postponed, and stalled before it is considered a "new" project, subject to more up-to-date interpretations of the law. In the coming years, the construction of Plant Scherer would be repeatedly delayed, postponed, and stalled, yet it was never forced to meet updated pollution regulations. The final unit did not come online until 1989, and when it did, the unit was roughly five times as dirty as a power plant that would have been built under 1989 law. What the case boiled down to was this: had Georgia Power entered into contracts simply to evade the scrubber requirement of the new law?

Georgia Power declined to allow me to speak to a lawyer or top executive about the specifics of the case. But a spokesperson for the company, who said her ability to comment was limited by the fact that the case was still unresolved, pointed out that knowledge and intent are not at issue in civil enforcement actions, and that the company believes it constructed Plant Scherer units 3 and 4 "in compliance with all applicable Clean Air Act requirements." In addition, the spokesperson said that prior to and during construction of Plant Scherer, the company worked closely with the Georgia Environmental Protection Division and the EPA on the permits for the plant.

That may be true, but a closer look at the history of Plant Scherer suggests that the story is a little more complicated.

- Georgia Power significantly changed the design of the power plant several times in the early years of construction. For example, the initial contract for the generators for units 3 and 4 was dated May

23, 1975. On November 4, 1975, that order was put on hold, pending a change in size and design. The size of the units was downscaled from 869 megawatts to 818 megawatts.

- The construction of the power plant was hardly one continuous project. Indeed, soon after construction at Plant Scherer supposedly began in 1975, it stopped. The *Georgia Power 1976 Annual Report* admits as much: "work was resumed in 1976 on three power plants where construction had been halted due to the company's financial problems in late 1974 and 1975 . . . About three million cubic yards of earth were moved at Plant Scherer, where work was resumed in May 1976." In 1979, Georgia Power's management decided to delay operation of units 3 and 4 by two years.

- Delays at Plant Scherer were not the result of weather or construction problems. As one internal Georgia Power report put it, "The commercial operation date changes are not a result of project-caused delays. The changes were directed by the Project Board as a result of economic considerations or load forecasts."

- It took fourteen years (1975–1989) to complete Plant Scherer, roughly twice as long as it took to build comparable power plants. Detroit Edison's Monroe power plant in Michigan, which is a similar size and uses similar technology, was built in seven years (1967–1975).

In addition, Georgia Power often argues that when Plant Scherer was built, scrubbers were still a new, untested technology. In fact, however, when the first unit at Plant Scherer went online in 1982, at least twenty-two power plants with various types of scrubbers were in operation around the United States. By 1989, when the last unit went online, there were literally hundreds of scrubbers in use.

In the end, whatever the legal technicalities may or may not have been, what the Southern Company essentially ended up with was a 1989 coal plant built to 1974 environmental standards. And in the electric power industry, that's pretty much like driving a Porsche in a race against a Hyundai.

With the election of President Bush in 2000, resolving the conflicts over New Source Review was one of Big Coal's top priorities. But politically, the industry had to be careful. Ordinary Americans were hardly clamoring for looser enforcement of the laws regulating coal-fired power plants. If anything, it was the opposite. A Gallup poll in 2001 found that 81 percent of Americans supported stronger environmental standards for industry. Another poll found that only 11 percent thought the government was doing "too much" to protect the environment.

Nevertheless, Big Coal was determined to do something about the lawsuits, which it viewed as a threat to its bottom line. Why spend hundreds of millions of dollars to clean up coal plants if it didn't absolutely have to?

For Big Coal, the timing couldn't have been better. The long, tortured energy crisis that began in California in the summer of 2000 and climaxed with rolling blackouts in January 2001 made electricity the talk of the nation. We know now that the so-called crisis was largely caused by market manipulation and regulatory breakdown, not by a shortage of supply, but we didn't know it then. The crisis worked greatly to the coal industry's advantage, giving President Bush the economic cover he needed to support its cause. As the president told CNN shortly after the 2001 blackouts, "If there's any environmental regulation that's preventing California from having 100 percent max output at their plants — as I understand there may be — then we need to relax those regulations."

Initially, Big Coal's main pressure point was Vice President Dick Cheney's energy task force, officially called the National Energy Policy Development Group, where a rollback of New Source Review was high on the "to do" list. And nobody was pushing harder for these rollbacks than Southern. A March 23, 2001, e-mail message from Mike Riith, a Southern Company lobbyist, to a Department of Energy official was titled "NSR and Energy Strategy" and included a memo that Southern had worked up about the problems with New Source Review. In the memo, Southern complained about the "extreme ways" the regulations were being interpreted by the EPA, which "threatens the safe, reliable, and efficient operation of energy production facilities across the country."

In case that wasn't enough, a pack of energy companies led by Southern formed the National Electric Reliability Coordinating Council and hired the most influential lobbyist in Washington at the time, Haley Barbour, to argue their case in a face-to-face meeting with Vice President Cheney. Barbour, who is now the governor of Mississippi, was one of Southern's favorite power brokers. Between 1998 and 2004, Southern paid his firm $1.56 million. Joining Barbour in his visit with Cheney was Bush's old friend Marc Racicot, the coal-friendly former governor of Montana, board member of the BNSF railroad, and lobbyist. (In 2004, Racicot became chairman of Bush's reelection campaign.) Scott Segal, a longtime spokesman for the electric power industry, said that Barbour and Racicot "did not dwell" on the lawsuits filed by the Department of Justice but suggested that the administration should abandon the standards that the Clinton administration had applied in bringing them.

Southern and its allies knew that gutting the existing regulations also would undermine the lawsuits: why pursue a violation of a law that is no longer on the books? This point was made explicit in a memo that EPA administrator Christie Whitman sent to Cheney after she got wind of the vice president's meeting with Barbour and Racicot: "As we discussed, the real issue for the industry is the enforcement cases. We will pay a terrible price if we undercut or walk away from the enforcement cases; it will be hard to refute the charge that we are deciding not to enforce the Clean Air Act." If nothing else, Whitman's arguments may have succeeded in forestalling any recommendation in the task force report — issued two weeks later, on May 17, 2001 — to rewrite the rules or cripple the lawsuits. Instead, the task force called only for the EPA to review the rules with the Energy Department, whose focus is to promote energy supply, and for the Justice Department to examine whether the suits were valid. Even if it wasn't a total victory for Big Coal, within the EPA the message was obvious: "We had an 'oh, shit' reaction," former EPA enforcement chief Eric Schaeffer told me. "It was clear that they were getting ready to spread around a lot of money to beat us with lobbyists."

Not surprisingly, many environmental groups went on the rampage, singling out the Southern Company for its backroom efforts to reinterpret the law. Full-page ads in the *Atlanta Journal-Constitution* slammed Southern as a "giant among polluters" and reminded readers that coal plants cut short the lives of thirty thousand people each year. Billboards went up with images of dirty smokestacks and slogans such as: "Southern Company: Bringing You Electricity and As Much Pollution a Year As 17 Million Cars." One report pointed out that just one big coal burner, Georgia Power's Plant Bowen in Bartow County, "emits as much sulfur dioxide as all the utility plants in the eight states of Utah, Delaware, South Dakota, Montana, Maine, Oregon, California, and Rhode Island combined." The same report pointed out that in 1999, Southern spent more than $4 million on its lobbying efforts, by far the most money spent by any company in the electric power industry.

All this had about as much impact on Southern Company executives as a gang of kids with slingshots would have on the crew of an M1 tank. "Inside the company, the view was, basically, 'Who cares what the enviros are saying?'" recalls one former Southern Company executive. "They were feeling besieged from all sides — deregulation, the discrimination lawsuit, and now this. They believed that the enviros were going to hate them no matter what they did, so why change? Why even try to placate them?" When this executive argued for the benefits of a proactive strategy on power plant pollution — why not get ahead of the fight, spend a few hundred million to clean up the plants, and reap the benefits of the positive PR? — he was met with blank stares. "In some ways, Southern is more like a feudal society than a modern company," he told me. "Their whole business model is built around denying change."

The fight over New Source Review continued for two more tortured years. Finally, on a sleepy Friday afternoon in August 2003 — a time deliberately chosen to sneak under the media radar — the EPA issued new regulations "clarifying" the New Source Review rule, allowing modifications of old plants as long as the cost of the repairs didn't exceed 20 percent of the total cost of the power plant. In fact, it was a clever way of repealing the regulations entirely, since on a $1 billion power plant, that would allow nearly $200 million in repairs or upgrades. (Think about it this way: how long could you keep a $25,000 car running if you could spend $5,000 a year on maintenance and repair? A long time.) Although the rollback would be challenged legally by environmental groups, companies such as Southern were happy to keep wrestling in court. When it comes to cleaning up the old coal plants, delay equals victory.

Southern didn't entirely thumb its nose at the rules. It's just that instead of following the strict interpretation of the law, the company cleaned up the plant in its own way and in its own time. In 1995, units 3 and 4 switched to low-sulfur Powder River basin coal, which reduced sulfur dioxide emissions. New, improved burners (the nozzles that inject the coal-air mixture into the boilers) were installed in 2001 and 2002, which cut nitrogen dioxide emissions. With these changes, Plant Scherer, circa 2004, was marginally cleaner than it had been when it was built. And in fairness to Southern, there are a number of old coal burners that are worse offenders than Scherer (the Hatfield's Ferry plant in Pennsylvania, for example, belches out sulfur dioxide at five times the rate of Plant Scherer). Nevertheless, by cleaning up the plant according to its own schedule, Southern bought itself twelve years of competitive advantage.

All in all, it's hard not to admire Southern's diabolical political skills. One of the many virtues of building a dirty power plant is that when the time comes to clean it up, you can boast about how much money you've spent and how much cleaner the air is. No one is better than Southern at touting the $2 billion it has spent over the past fifteen years to clean up its old coal burners and how the total emissions from its power plants are dropping. Those are, of course, good things. What is usually overlooked, however, is the fact that it is only because the power plants were so dirty to begin with that Southern is able to claim such progress. If you view the issue narrowly, ignoring the premature deaths and asthmatic children, you can make the case that this is just smart business. It gave Southern years of fat profits, and when the time came to invest in pollution controls, they could boast about being good corporate citizens. Who doesn't have a soft spot for a reformed sinner?

And you can't argue that Southern doesn't take care of its friends. Just a week or so after the rollback of New Source Review was announced in August 2003, the *Atlanta Journal-Constitution* reported that Southern had hired John Pemberton, the chief of staff to Jeffrey Holmstead, the EPA's air policy guru, as a lobbyist. Holmstead was one of the main architects of the rollback. As his top aide, Pemberton was widely regarded as having been involved in every detail. Pemberton assured reporters that he had recused himself from the issue when he began negotiating with Southern. But few outsiders bought it, in part because Pemberton's recusal had covered only the last few weeks before his departure. The EPA had formally proposed the new rule seven months earlier, and the administration had been actively trying to resolve the issue practically from the moment President Bush took office. "We of course do not leap to judgment," Charles Peters wrote in the *Washington Monthly*. "But cynics might conclude that the job is a payoff for the rule."

PART III

THE HEAT

Chapter 8

Reversal of Fortune

IN RECENT YEARS, more words have been written about the dangers we face from a superheated climate than almost any subject other than terrorism. Melting icecaps in the Arctic, retreating glaciers in China and Tibet, droughts in southern Africa, vanishing islands in the Pacific Ocean, the deaths of twenty thousand people in Europe from an unexpected heat wave — even a short list of the effects of a warming climate quickly becomes overwhelming. And most of us have participated in barstool conversations about weird weather — the unseasonable droughts, the hot summers, the big rains. The weird events up for discussion may vary in different regions of the world, but the subtext is always the same: is it just me, or is something strange going on here?

This question took on sudden urgency in the late summer of 2005 when Hurricane Katrina slammed into the Gulf Coast, killing more than 1,700 people in seven states, causing an estimated $80 billion in property damage, and nearly wiping a great American city off the map. Several hurricane experts noted that the storm's particular intensity may well have been due to the warming waters of the Gulf of Mexico, which helped to rev up the hurricane's engine. Certainly, Katrina was a natural disaster, but was it also a preview of what life might be like on Planet Greenhouse?

Still, the subject of global warming remains beyond our grasp. In part this is because of our narrow perspective. Global warming is not an event that is occurring on a human scale. It is about climate, not weather, although the two are obviously related. In the earth's 4.5-billion-year history, the climate has gyrated wildly, from blast furnace conditions hot enough to melt steel to deep ice ages lasting millions of years. And although a sudden warming of 5 to 10 degrees Fahrenheit would have a huge impact on civilization as we know it, compared to the tumultuous changes in the past, it barely qualifies as a perturbation. The commonly held view of the earth as a place with a steady, stable, nurturing climate is an entirely human invention, a manifestation of the fact that civilization emerged during a pleasant and stable ten-thousand-year period known as the Holocene, in which we are still living today. But in the context of the earth's long history, these temperate years are an exception, not the rule. Human beings simply have not been around long enough to witness Mother Nature on a rampage.

There is also the problem of complexity. It's one thing to understand that increasing concentrations of greenhouse gases such as carbon dioxide (CO_2) are contributing to the gradual warming of the earth. It's quite another to understand the ways in which melting ice in the Arctic influences drought patterns in Africa, or how higher evaporation rates lead to increased cloud cover, which creates slightly warmer nights and cooler days in certain regions of Central and South America, encouraging the growth of a deadly fungus that is killing off dozens of species of harlequin frogs. And even if you can grasp the big picture, connecting it to what happens behind the light switch is another matter entirely. The complexity of what is going on here is so great, and the interconnections so subtle and profound, that anyone who tries to come to grips with it is inevitably drawn back to the old metaphor of the butterfly flapping its wings in China and causing a hurricane in Florida. "This is one of the big dilemmas for scientists," Lloyd Keigwin, a paleoclimatologist at the Woods Hole Oceanographic Institution, told me during a research trip I took with him into the North Atlantic. "How do we communicate what's at risk? How do we communicate the fact that, on earth, everything is connected?"

The success of the coal industry is largely dependent on our failure to make that connection. Big Coal's goal is to keep us comfortable, not curious. It's not hard to understand why. Coal is by far the most carbon-intensive of all fossil fuels, emitting more than twice as much CO_2 per unit of energy as natural gas, and so any limits on CO_2 emissions will hit coal the hardest. In addition, if the world ever gets serious about tackling global warming, coal-fired power plants are a much more likely target for a crackdown than, say, SUVs, which are beloved by millions of American voters. "We're walking around with a big bull's-eye on our foreheads," jokes James Rogers, the CEO of Duke Energy, one of the largest coal burners in America. Also, CO_2 isn't like other pollutants, which can be removed from an old coal burner by bolting on a scrubber. With current technology, the most economical way to cut CO_2 emissions from an old plant is usually to shut it down. And unlike other carbon-intensive industries such as steelmaking, moving to China is not an option for coal-fired power generators.

Big Coal's empire of denial is not monolithic. Back in 2001, when I interviewed Rogers for the *New York Times Magazine*, he told me that he had little doubt about the scientific reality of global warming and that he supported limits on CO_2 emissions. To Rogers, limits were inevitable in the near future: the sooner the industry admitted that fact, the better. Paul Anderson, chairman of Duke Energy's board of directors, has called global warming "one of the most pressing issues of our time" and supports a national tax on CO_2 emissions. When asked about the difficulty of capping CO_2 emissions, Wayne Brunetti, CEO of Xcel Energy, another big coal burner, was remarkably blunt: "Give us a date, tell us how much we need to cut, give us the flexibility to meet the goals, and we'll get it done." While researching this book, I met dozens of coal miners, railroaders, and power plant engineers who had no doubt that burning coal was heating up the planet and who regretted their roles in keeping the fires burning. "But I have a family to feed," one Southern Company employee told me as we stood beside a mountain of coal at Plant Scherer. "I don't have the luxury of being moral about it."

Such frankness is the exception, not the rule. For the most part, Big Coal has kept up a unified front. The party line has evolved slightly, moving from outright denial to a risk-management strategy that acknowledges the earth is getting warmer and humans might be responsible, but argues that (A) the consequences will be, on the whole, modest and (B) we have plenty of time to work it out. For Big Coal, argument B is key. Off the record, many coal industry executives and consultants will tell you that people in the industry are divided between those who believe that global warming is a hoax and those who acknowledge that the science is pretty compelling. But there is one thing both sides agree on: delay is good. The longer Big Coal can draw out this fight, convincing us — and, more important, their political allies in Washington — that this is a problem we should deal with in 2050, not 2010, the better off it will be. Perhaps, the industry argues, Craig Venter, the scientist who sequenced DNA, will build CO_2-eating bacteria that will solve all our problems, or someone will figure out a way to put solar panels on the moon and beam endless energy back via microwaves. So why worry? We'll solve this problem later.

For Big Coal, delay has turned out to be a very profitable strategy. For the rest of us, the benefits aren't so clear.

Unlike some of the nasty pollutants that coal plants pump into the atmosphere, CO_2 is tame and friendly. It's an invisible gas made up of one molecule of carbon and two molecules of oxygen. It puts the fizz in Coke and causes bread to rise. In solid form, it's dry ice. One way to think about the earth is as a giant carbon-breathing organism, with CO_2 constantly being exhaled by humans and other animals, geologic sources such as volcanoes, decaying plant and animal matter, and burning fossil fuels, then being inhaled by billions of growing plants, which use a process called photosynthesis to steal the energy in the carbon molecule and release oxygen. Vast amounts of CO_2 are also absorbed into the oceans, where tiny marine creatures use it to build calcium carbonate shells. When those creatures die, their shells eventually settle to the bottom of the sea and form limestone sediments. In the atmosphere, CO_2 functions like the glass panes in a greenhouse: it lets short-wave solar radiation in but doesn't let much long-wave radiation (heat) out. Without the protecting insulation of greenhouse gases, the earth would be a frigid, uninhabitable place. On Venus, where CO_2 levels are much higher, the planet's surface is hot enough to melt lead.

This carbon cycle is a naturally balancing system, with the amount released roughly equal to the amount absorbed. But by burning fossil fuels, which are by definition carbon based, we have begun to alter that balance. As we pump more and more CO_2 into the air, the earth can no longer absorb it, and so it begins to build up in the atmosphere, trapping more and more heat and warming up the planet. There are more than twenty different greenhouse gases, including methane, which is less common than CO_2 but more potent. Other greenhouse gases are less of an immediate issue than CO_2 because industrial society is not dumping hundreds of millions of tons of them into the atmosphere each year. But as the earth warms, it creates a positive feedback cycle that could change that. Recently, scientists have become concerned that melting permafrost beneath west Siberia could free billions of tons of methane from peat bogs. Some climatologists fear that a sudden methane release of this magnitude could radically destabilize the earth's climate.

The basic physics of the greenhouse effect are not complex and have been more or less accepted by the scientific establishment for more than one hundred years, since the Swedish scientist Svante Arrhenius scrawled a few calculations on the back of an envelope in 1896. Also beyond question is the fact that since the world began burning fossil fuels, the level of CO_2 in the atmosphere has climbed from about 280 parts per million in 1800 to 380 parts per million today. It is now rising at a rate of about 1.5 parts per million each year. Over the past century, the average temperature on the earth's surface has risen approximately 1 degree Fahrenheit (0.6 degree Celsius), with most of the increase occurring in the past two decades. By 2100, unless something changes, the earth's temperature is expected to increase by an additional 2 to 10 degrees Fahrenheit (1.5 to 5.8 degrees Celsius). An increase at the upper end of this range would have a dramatic impact on life as we know it, including severe droughts and rapidly rising sea levels.

The real complexity of global warming is not in the science, however, but in the politics. The battle began in 1988 with the dramatic testimony of James Hansen, a NASA scientist and one of the pioneers of global warming research, before the U.S. Senate. "The greenhouse effect has been detected," Hansen stated bluntly in his

CO₂ LEVELS IN THE ATMOSPHERE

Mauna Loa Observatory, Hawaii
Monthly Average Carbon Dioxide Concentration

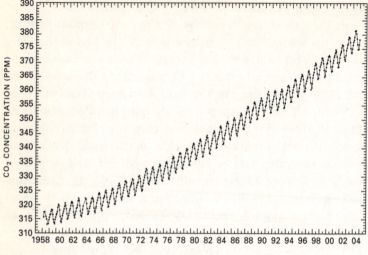

testimony, "and it is changing our climate *now*." To many scientists, this was not really news. In 1977, a report on global warming from the National Academy of Sciences warned of "highly adverse consequences" to the earth's climate if we continued to burn coal as an energy source. But Hansen's testimony, which came during a summer of heat waves and drought, made headlines. "I've never seen an environmental issue mature so quickly, shifting from science to the policy realm almost overnight," Michael Oppenheimer, an atmospheric scientist with the Environmental Defense Fund, told the *New York Times*. "It took a government forum during a drought and a heat wave and one scientist with guts to say, 'Yes, it looks like it has begun and we've detected it.' He felt comfortable saying clearly and loudly what others were saying privately. That's mighty important in the public policy business."

During the 1988 presidential campaign, George H. W. Bush, running against Democrat Michael Dukakis, used the heightened awareness of global warming to back up his credentials with environmentally conscious middle-class voters. "Those who think we are powerless to do anything about the greenhouse effect forget about the 'White House effect'; as President, I intend to do something about it," he said in Michigan on August 31, 1988. If he was elected, Bush promised, he would convene an international conference on the environment. "We will talk about global warming," he vowed, "and we will act."

But Bush didn't act. Shortly before he took office, the United Nations Intergovernmental Panel on Climate Change was established by two UN organizations, the World Meteorological Organization and the United Nations Environment Programme. Its purpose was to assess the "risk of human-induced climate change." The United States was happy to be among the thirty-five nations assembling a state-of-the-art report on the science and impacts of, and responses to, global warming. But when it came to forging a pact to actually do something, that was a different story. Shortly after Bush took office, his chief of staff, John Sununu, spearheaded efforts to stall action and dilute the congressional testimony of leading scientists such as Hansen about the seriousness of global warming. In fairness to Bush, who often called himself "the environmental president," he did take action on what was then the most pressing pollution problem, acid rain, helping to push through the 1990 amendments to the Clean Air Act, which included the innovative sulfur dioxide trading program (described in chapter 6) that dramatically reduced pollution over the coming years. When it came to fighting Big Coal, however, that was as much political capital as he was willing to expend.

During the next few years, during international negotiations over greenhouse gas reductions under what was known as the United Nations Framework Convention on Climate Change, the Bush administration's position was summed up by one commentator as "disinterest and opposition." But to William Nitze, a high-ranking U.S. delegate in the UN-sponsored climate negotiations, it was more than that: Sununu's political opposition to the agreement was influenced by the power of Big Coal in states that President Bush would need for reelection. In part because of this, the convention treaty, which was signed at the Earth Summit in Rio de Janeiro in

1992 and laid the groundwork for the Kyoto Protocol, was weaker than it might have been and failed to resolve tensions between rich and poor nations over emissions cuts. These same flaws would haunt Kyoto. Still, the Rio agreement — which virtually every major nation, including the United States, signed and ratified — committed the world to a common goal: "stabilization of greenhouse gas concentrations in the atmosphere at a level that would prevent dangerous anthropogenic interference with the climate system."

The Rio treaty was a mighty blow for Big Coal. In the United States, many coal and power companies were used to having their way with state and federal regulators. Now they had an international army of scientists and bureaucrats marching after them, and they weren't happy about it. At a worldwide coal industry conference in Barcelona in 1993, Dr. Harlan Watson, who had been a member of the Bush administration's team in climate talks and who has, over the years, proven himself to be a solid ally of Big Coal and Big Oil, warned the industry about the implications of the Rio agreement and the coming crackdown on CO_2 emissions. "You need to take this matter very, very seriously," he told coal industry executives.

And take it seriously they did. In *The Heat Is On,* veteran journalist Ross Gelbspan's account of the early days of the fossil fuel industry's fight against global warming, Gelbspan offers a blow-by-blow account of a "mystification campaign" that made extensive use of a small group of scientists whose views contradicted the consensus of the world's experts and that promoted those views to a gullible public. "It lulled people into deep apathy about the crisis by persuading them that the issue of climate change is terminally stuck in scientific uncertainty," Gelbspan wrote in 1998. "It is not."

One of the most prominent industry groups in the early years of the global warming debate was the Western Fuels Association, a group of coal-burning electric utilities and cooperatives. In the early 1990s, when the battle over global warming began to be engaged, Western Fuels began a campaign to undermine the science of climate change. In many respects, it was the same strategy the industry had used for decades to question the health impacts of air pollution. "There has been a close to universal impulse in the [fossil fuel] trade association community here in Washington to concede the scientific premise of global warming . . . while arguing over policy prescriptions that would be less disruptive to our economy," the group admitted in its 1994 annual report. "We have disagreed, and do disagree, with that strategy." Western Fuels was one of the key backers of the Information Council on the Environment, a front group whose explicit purpose, according to Gelbspan, was to "reposition global warming as theory rather than fact."

To help do that, Western Fuels funded a handful of global warming skeptics to travel around the country and speak to the press and the public. The campaign targeted "older, less educated males" and "younger, low-income women" in districts that received their electricity from coal and that preferably had a representative on the House Energy and Commerce Committee. Western Fuels also spent $250,000 to finance a video called *The Greening of Planet Earth,* which aimed to convince policymakers that a warmer, wetter earth would be, contrary to the alarmism of environmentalists, a lush, lovely place. The video is a modern classic of industrial PR, full of shots of dense forests and fertile fields, suggesting that burning coal is a good way of fertilizing the world's flora. (Higher CO_2 levels in the atmosphere do stimulate growth in many trees and plants, but only briefly. Other factors, such as nutrients in the soil, soon limit growth.)

"The Western Fuels campaign was extraordinarily successful," Gelbspan wrote. "In a *Newsweek* poll conducted in 1991, before the spin began, 35 percent of respondents said they 'worry a great deal' about global warming. By 1997 that figure had dropped by one-third, to 22 percent."

Perhaps one reason the Western Fuels campaign was so successful was that it was guided by a true believer. During most of the 1990s, both the Western Fuels Association and the Greening Earth Society, a PR group formed by the association to promote the view that burning coal would lead to a healthier, happier planet, were headed by a man named Fred Palmer. Today Palmer is an executive vice president at Peabody Energy, the world's largest coal company. In the early 1990s, he was one of the most outspoken promoters of Big Coal's argument that higher levels of CO_2 would lead to a greener, more productive world. For Palmer, fossil fuels are, quite literally, a gift from God. "It is easy to conclude that, under a preordained plan, coal and oil lay in wait for exploitation by humans to permit our creation of an environment on Earth conducive to the spectacular success of our species," Palmer told an international coal conference in 1996. He argued that the 1992 United Nations Framework Convention on Climate Change was based on a vision of "apocalypse, scarcity, drought, famine, and pestilence" and that the agreement was designed "to limit the material progress of the human community based on a moral view of the present and future that by definition rejects the pro-human vision of all the world's great religions." To Palmer, taking action on global warming was not so much a question of science as it was a question of faith. Why would the world have so much coal if we weren't meant to burn it?

But as even Palmer would admit, faith was no substitute for good, solid numbers. In every debate about limiting CO_2 emissions, the central question was always this: what will the economic impact be? For this, economists had to rely on computer models that were at least as complex as the models used by climate scientists, but without being held accountable to any peer review process. According to Jason Shogren, a professor at the University of Wyoming who was also the senior economist for environmental policy on the president's Council of Economic Advisers during the run-up to Kyoto, credible estimates of the treaty's economic impact ranged from 0.5 to 2.4 percent of the U.S. gross domestic product (GDP) by 2010. In Shogren's view, the best numbers came from Stanford University's Energy Modeling Forum, which compared a diverse group of economic models employing different parameters. This study, which was cosponsored by the Electric Power Research Institute, estimated that if the benefits of a global carbon trading program were included in the treaty (similar to the market for sulfur dioxide emissions that was created by the 1990 amendments to the Clean Air Act) the economic cost to the United States would be between 0.3 and 0.5 percent of GDP, or $30 billion to $50 billion per year.

Not surprisingly, the coal industry's own estimates were much higher. In congressional testimony in 1998, Ohio coal operator Bob Murray cited a study by an industry-friendly research group that es-

timated Kyoto would cost the United States 3.2 percent of GDP, or
$300 billion, by 2010. Peabody Energy went further, using eco-
nomic scare tactics that put Murray to shame. During the debate,
the company pushed a study by a now defunct research group in
Boulder, Colorado, that warned that if America signed on to Kyoto,
$1.8 trillion, or 19 percent of GDP, would be "at risk" in 2010. To put
it another way, Peabody's estimate was forty to fifty times higher
than estimates from the Energy Modeling Forum. In another report, Peabody claimed that Kyoto would increase the typical household electric bill by 82 percent, force a tax on coal of $109 per ton,
and cost the nation a million jobs.

And where would all these jobs go? To China, India, and other developing nations, all of which, the coal industry pointed out, would
be exempt from binding restrictions on CO_2 emissions. Thus, cutting CO_2 was effectively linked with the decline of manufacturing
jobs, economic insecurity in the Midwest, and deep-seated worry
about the rising power of Asia.

In July 1997, as the date for the Kyoto meeting approached, West
Virginia senator Robert Byrd, a passionate and powerful advocate
for Big Coal, cosponsored a Senate resolution that expressed the
opinion that the United States should not sign any agreement in
Kyoto unless it included commitments by developing nations such
as India and China to cut emissions or if it was deemed to "result in
serious harm to the economy of the United States." The resolution
passed the Senate by a vote of 95–0. The message to climate negotiators was clear: you can talk about global warming all you want, but
it will be a cold day in hell before America agrees to any serious action to address it.

In December 1997, 6,000 delegates from 160 countries met in
Kyoto, Japan, to hammer out the agreement that would become
known as the Kyoto Protocol. It's inaccurate to suggest that the
United States was the Big Bad Wolf during the long and exhaustive
negotiations: many Europeans seemed more interested in economically punishing the United States than in achieving a workable plan
to reduce greenhouse gas emissions, steadfastly refusing to include
any mechanism (such as allowing energy companies to meet their
emissions targets by investing in carbon-absorbing projects like re-

forestation) that would give energy companies flexibility in how or when they achieved their emissions reductions. "There was clearly a moral edge to the Europeans' position," one American negotiator who was in Kyoto told me. "To them, America's consumption of fossil fuels was downright sinful, and they wanted the agreement to hold us accountable for that."

The United States, on the other hand, put its own economic interests above all else, including the welfare of poor nations threatened by global warming. Holding the treaty hostage because China and India were not party to the agreement — a point that had been underscored in the U.S. Senate resolution — was an especially shameful negotiating tactic. It is indeed true that these nations emit large amounts of CO_2 (and will emit even more in years to come), but America, as a nation that had grown rich by burning fossil fuels, and that was, as much as anyone, responsible for overloading the atmosphere with CO_2, certainly had a moral obligation to show leadership in dealing with the problem of global warming.

In the end, Vice President Al Gore flew to Kyoto at the last minute to help broker a deal that committed the United States to cutting greenhouse gas emissions by 7 percent (based on 1990 levels) between 2008 and 2012. The European Union agreed to an 8 percent reduction, Japan to 6 percent, and twenty other nations to 5.2 percent. In November 1998, the U.S. ambassador to the United Nations signed the treaty on behalf of the Clinton administration, but it was largely a ceremonial gesture. By that time, Clinton was distracted by the rising conflict in Kosovo. And thanks to the investigation into his affair with White House intern Monica Lewinsky, his political capital was at an all-time low, leaving him in no position to provoke a titanic battle with Republicans over global warming. Not surprisingly, Clinton never submitted the treaty to Congress for ratification.

"The 2000 election was a big one for us," Jack Gerard, the head of the National Mining Association, the coal industry's powerful trade and lobbying group, told me as we sat in his office in Washington, D.C., the following summer. That was an understatement. Despite the successful scuttling of U.S. participation in the Kyoto Protocol, global warming was too big and too urgent an issue to simply melt into the background. Indeed, many people in the coal industry saw the election as yet another make-or-break moment for the future of coal. Democratic presidential candidate Al Gore was not so politically naive as to be openly antagonistic to Big Coal. He supported funding to develop "clean coal" technologies, and his labor-friendly views won him the support of the United Mine Workers union. But coal industry executives knew what was in his heart, and they knew it wasn't coal.

George W. Bush, who needed to win moderate voters, went out of his way to demonstrate that Gore was not the only candidate concerned about the fate of the planet. "Global warming needs to be taken very seriously," he said during one debate. He made it clear that he supported regulating CO_2 and agreed that limits should be set on emissions (although he also stated that he didn't back the Kyoto Protocol). This view was repeated in an energy policy paper the Bush campaign released in late September 2000. In it, Bush had the moxie to try to "out-Gore" his opponent, arguing that he supported mandatory limits on CO_2, while "Vice President Gore has advocated only a voluntary program."

Nobody in the coal industry took Bush's campaign rhetoric seriously. Whatever he said or did not say during the campaign, they knew he was a fossil fuels man to the core. By contrast, Gore could talk about clean coal technology all he wanted, but Big Coal executives knew that if he was elected, he would install solar panels on the White House roof and hold press conferences on sinking islands in the Pacific. Nevertheless, when the votes were tallied, it was a tight race even in coal country: Bush won Kentucky, Ohio, West Virginia, Wyoming, and Gore's home state of Tennessee. But Gore won two important Big Coal states, Illinois and Pennsylvania.

When Bush was declared the victor after the disputed Florida recount, most outsiders presumed that he would make good on his campaign promise to regulate CO_2 emissions. He had, after all, campaigned on a theme of integrity. In addition, one of Bush's biggest campaign contributors was Ken Lay, CEO of Enron, the energy trading company that would soon implode. Lay was an advocate of regulating CO_2 not because he wanted to save the planet, but because it would help accelerate the shift to natural gas, a much less carbon-intensive fuel. "We knew Bush was on our side," one coal industry executive told me. "But frankly, some of us were concerned."

There was also concern about Christie Whitman. As EPA administrator, Whitman saw herself more in the tradition of Theodore Roosevelt than Ronald Reagan, an old-fashioned Republican who believed that conservation was a conservative principle. Indeed, many of America's landmark environmental laws had been passed by Republicans: the Clean Air Act (Nixon), the Endangered Species Act (Nixon), the Toxic Substances Control Act (Ford), and the acid rain amendments to the Clean Air Act (George H. W. Bush). Whitman's mistake was believing that President George W. Bush saw himself the same way.

In fact, Whitman didn't know the president very well. Back when they were both governors of major states, they had talked at a few conferences, and she had a general sense that she was on the same page with him on environmental issues. But they never had any substantial conversations about it, either during their brief meeting to discuss the EPA job or afterward. Nevertheless, at her Senate confirmation hearing, she boldly proclaimed that America was ready "to enter a new era of environmental policy — an era that requires a new philosophy of public stewardship and personal responsibility."

A few days before President Bush's inaugural, as the new president was assembling his team of transition advisers, the *Washington Post* reported that Big Coal was "particularly well-represented on the Bush team." And Peabody Energy was particularly prominent among the Big Coal representatives. Irl Engelhardt, the company's aggressive fifty-three-year-old chairman and CEO, was appointed to the EPA transition team. Peabody vice president John Wootten and Steven Chancellor, chief executive of a Peabody subsidiary, Black Beauty Coal, were on the Energy Department team. And Thomas Sansonetti, a former lobbyist with a Wyoming law firm representing Peabody and other energy firms on leasing matters, headed the group choosing the top personnel for the Interior Department.

How did Peabody score such access? It didn't hurt that in the 2000 election, Peabody's holding company gave $846,000 to federal campaigns, 98 percent of which went to the Republican Party. That was by far the largest contribution by a coal company, and about five times as much as Peabody's rival, Arch Coal. To put these numbers in perspective, the biggest contributor among renewable energy companies was the American Wind Energy Association, which gave a whopping $34,600, only 40 percent of which went to the GOP.

The money was well spent, because no coal company had as much at stake in the 2000 election as Peabody. Peabody Coal, as the company was originally called, was named after its founder, Francis S. Peabody, an early-twentieth-century coal baron who had become fabulously wealthy by providing coal for Samuel Insull's power plants in the Midwest. The company had thrived and expanded over the years, with particularly large holdings in the Powder River basin. For most of the year 2000, Engelhardt had been quietly putting together a plan to take the company public. This move could potentially generate millions of dollars in cash that would help the company fund new ventures, including an expansion into China, as well as to grow beyond coal mining and move into the power generation business. All this depended on a number of factors, including the election of George W. Bush and a coal-friendly political climate.

Engelhardt once called his life "an all-American story." He grew up in Pinckneyville, Illinois, an old coal town near the southern tip of the state. When Engelhardt attended high school there in the early 1960s, the strip mines were booming. Today the mines are mostly gone, and like most old mining towns, Pinckneyville is a bleak place, scarred and desolate-feeling. One of the biggest employers in the region is a prison. Engelhardt, who grew up in a small house outside town near the state highway, did not have an easy time of it. His father was killed in a truck accident when he was five years old, leaving his mother to raise him and his three siblings on her own. After high school, he attended the University of Illinois at Urbana-Champaign, then took a job at a small accounting firm, specializing in tax and audit issues. A few years later, he earned an MBA from Southern Illinois University, then went to work as a consultant for Arthur Andersen, the prominent accounting firm that would later be one of Enron's primary enablers. He joined Peabody in 1979 and began working his way up the corporate ladder, becoming CEO in 1990.

Less than two weeks after Bush took the oath of office, Vice President Dick Cheney went to work on what seemed to be one of the administration's top priorities: paying back its pals in the energy business. This goal was carried out under the guise of Cheney's energy task force, which was charged with coming up with a new energy policy for America. The task force was made up of many of the top officials in the new administration, including Larry Lindsey, an economic adviser to President Bush; Spencer Abraham, the energy secretary; Paul O'Neill, the Treasury secretary; and Christie Whitman. It held its first meeting in the vice president's ornate Ceremonial Office on February 9, 2001. Although there were no industry representatives on the task force, they were camped right outside the room and knew very well what was going on inside. Exactly what was discussed in these meetings has never been disclosed. The Bush administration, citing executive privilege, has fought requests by Congress and others for the minutes of the meetings all the way to the Supreme Court.

Other coal and power companies were content to use their influence with the new administration to press for regulatory relief on issues such as New Source Review, but Peabody had bigger plans. Exactly three days after the task force met for the first time, Peabody notified the Securities and Exchange Commission (SEC) that it intended to take the company public in the coming months. It was a bold strategy — the market was still in the doldrums after the Silicon Valley meltdown of 2000, and coal companies had not traditionally been viewed as big growth stocks. But Engelhardt and others at Peabody knew better. They knew that America's appetite for energy is a far more powerful force than any concern for global warming, and that to satisfy that appetite, people would inevitably turn to their last great cache of fossil fuel: coal. In addition, they knew that Dick Cheney was a man they could rely on, and they believed that, with Cheney's help, the Bush administration would open the door to an era of unprecedented growth and opportunity for the coal industry. And just to show how serious they were, on the same day the company filed the paperwork to go public, it also announced plans to build a new 1,500-megawatt coal-fired power plant in Muhlenberg County, Kentucky. It was the first major new coal plant announced in a generation. Lest anyone had any doubt, the coal boom was on.

One of the biggest misconceptions about global warming is that rising greenhouse gas levels will result in a slow, steady increase in the earth's temperature, as if the planet were simply being swaddled in an ever thicker blanket of CO_2. If that were the case, it would be easy to make the argument that global warming is a problem we have plenty of time to work out, and that a degree or two of warming — as Dr. Fred Singer, a noted global warming skeptic, put it at one conference I attended in 2004 — might lead to improved growing conditions in many agricultural regions and better beach weather in chilly climates. This is a comforting notion, and one that has long been promoted by skeptics like Singer. Unfortunately, it is wrong.

Until the 1960s, the earth was viewed by most geologists and climatologists as a stable, fairly inert place. The sun rose and set, ice ages moved in and out at 100,000-year intervals, and the earth's climate was a gentle, predictable system that worked in harmony with distant solar rhythms. Many geologists assumed that the present was the key to the past — that the forces at work over time were steady and constant, and that you could understand what happened a million years ago simply by extrapolating backward over time.

In the late 1960s, that vision collapsed. The earth's continents are not fixed islands, as had long been supposed, but remnants of a single landmass that broke apart about 225 million years ago. Furthermore, the outer surface of the earth is built of fractured plates, and those plates are constantly moving, bumping into one another, collapsing, breaking apart, and folding into great mountains — all this happening too slowly for us to notice. Similar breakthroughs were made in understanding the history of the earth's climate. Evidence discovered in ice cores and ocean sediments revealed that the earth's atmosphere was not a predictable system that moved in and out of ice ages with rhythmic grace, but a tempestuous beast that is capable of dramatic, unpredictable shifts in surprisingly brief periods of time.

Exhibit A: the Younger Dryas, a climatic event that ended 11,500 years ago, just as the climate was emerging from the last ice age. ("Dryas" refers to an alpine wildflower, *Dryas octopetala*, whose pollen was found in a Swedish bog containing sediments from that era.) After several thousand years of steady warming, temperatures in northern latitudes inexplicably plunged 10 to 15 degrees Fahrenheit and stayed that way for a thousand years, before warming up again just as suddenly. During the Younger Dryas, conditions in the United Kingdom and Europe were similar to those in the Arctic today. Icebergs drifted as far south as Portugal. By the 1970s, evidence of the Younger Dryas had been found in ocean sediments, ice cores, and other geologic features. Its existence was beyond dispute, but what had caused this abrupt climate change? Many climatologists believed the answer was to be found in the oceans, the great arbiter of the earth's climate, but what the exact mechanism was, and how it worked, was a mystery.

In the late 1980s, an ambitious geochemist at Columbia University's Lamont-Doherty Earth Observatory named Wally Broecker hypothesized that the abrupt changes of the Younger Dryas had been caused by a disruption of what he called "the great ocean conveyor." This conveyor, known to oceanographers as the thermo-haline circulation system, is a kind of global current driven by differences in temperature and salinity in the earth's oceans ("thermo" = heat; "haline" = salt). The North Atlantic is an important cog in this circulatory pattern. It's the spot where heat from the warm, saltier water flowing north in the Gulf Stream is released into the atmosphere. As the water cools, it becomes more dense, sinks to the ocean depths, and begins flowing southward. It is largely because of the heat given off by the Gulf Stream that Great Britain and much of northeastern Europe enjoy a relatively mild climate, despite their high latitudes.

Broecker hypothesized that the Younger Dryas may have been triggered when a giant lake of glacial meltwater — known as Lake Agassiz, which was located in what is now Manitoba, Canada — drained rapidly through the St. Lawrence River. As this sudden rush of fresh water — roughly thirty thousand tons a second — poured into the North Atlantic, it destabilized the conveyor at a crucial point in the system. Because fresh water is less dense than salt water, the cool diluted water tended to float instead of sink, effectively shutting down the mechanism that drives ocean circulation and brings heat to the northern latitudes. Thus, the gradual warming that had melted the ice sheets had the paradoxical effect of plunging the region into a sudden cold state that lasted nearly a thousand years. Broecker's theory is still controversial, especially the part about the meltwater flowing through the St. Lawrence. But for the first time, a mechanism had been identified that could begin to explain the abrupt changes in climate that perplexed scientists.

The idea of the earth's climate as a stable, balanced system was killed off for good in the early 1990s by new evidence discovered in Greenland ice cores. By examining the decay of carbon isotope ratios in air bubbles trapped in ancient ice, a team of researchers that included Pennsylvania State University paleoclimatologist Richard Alley determined that at the end of the Younger Dryas, not only had the earth's temperature warmed 15 degrees Fahrenheit, but it had done so in less than ten years. In geologic terms, ten years is virtually overnight. In human terms, it was like going to sleep one night in Alaska and waking up in Costa Rica. In the Greenland ice cores, paleoclimatologists found evidence of another climate jump about 8,500 years ago. The so-called 8.5K event was shorter and less dramatic than the Younger Dryas, but it was more proof that the earth's climate was nowhere near as stable as scientists had once thought. In fact, the earth's climate seemed to be a highly complex system that, based on small forces that were still dimly understood, tended to lurch from one steady state to another. "You might think of the climate as a drunk," Alley later explained. "When left alone, it sits; when forced to move, it staggers."

The implications of this discovery are enormous. It means that the real danger of continuing to pump millions of tons of CO_2 into the atmosphere is not that the earth's climate will warm steadily like water on a stove, but that our two-hundred-year fossil–fuel burning party could force the drunk out of his chair. If that happens, the old man could lead us right back to an event like the Younger Dryas.

From the beginning, global warming was at the top of Christie Whitman's agenda. Six weeks after she joined the EPA, she was scheduled to travel to Trieste, Italy, for a meeting with her G8 counterparts — the environmental ministers from Canada, Britain, France, Germany, Japan, Italy, and Russia. The official purpose of the meeting was to advance discussions on addressing climate change. "I was keenly aware that this preparatory meeting was the first opportunity for our closest allies to take the measure of President Bush's stance on environmental policies," Whitman later wrote. She understood that Bush was not about to support the Kyoto Protocol, but she believed that there might be other ways to move forward, especially given his campaign promise to support restrictions on CO_2 emissions.

Whitman knew very well what a complex and politically explosive issue global warming was, but she also knew that Bush's campaign pledge had been unambiguous. And despite her political savvy, she still believed that such pledges meant something. After all, Bush's mandatory cap on CO_2 emissions was included in a thick notebook titled "Transition 2001," the official collection of his campaign promises that had been written by the Bush-Cheney transition team. Whitman had received a copy of the collection when she had been nominated for the EPA job. But just to be sure there was no misunderstanding, before Whitman left for Trieste, she met with Condoleezza Rice, the president's national security advisor. She made sure that Rice knew she would be touting the president's campaign commitment to a mandatory cap on CO_2 emissions. According to Whitman, Rice agreed that it would be a sound approach. Whitman also checked with Andrew Card, the White House chief of staff, and again she got the green light.

Before she left, Whitman appeared on the CNN political talk show *Crossfire*. "George Bush was very clear during the course of the campaign that he believes in a multi-pollutant strategy, and that includes CO_2," she said in response to a comment from Robert Novak, the show's conservative commentator. "He has also been very clear that the science is good on global warming," she added, noting that "introducing CO_2 to the discussion" is a crucial step in attacking a genuine problem. The next day, which happened to be the morning of President Bush's State of the Union address, Whitman met briefly with reporters on Capitol Hill. "There's no question but that global warming is a real phenomenon that is occurring," Whitman said. "And while scientists can't predict where the droughts will occur, where the flooding will occur precisely, or when, we know those things will occur." In an interview with *Time* magazine, she went even further, ratcheting up the urgency and, for the first time, linking global warming with human activity — that is, the burning of fossil fuels. "The climate is changing more rapidly than we've seen in the past," she told *Time*, "and there are human actions that are contributing to what we're seeing."

All over Washington, D.C., in the offices of conservative representatives and politically connected think tanks such as the Competitive Enterprise Institute, alarm bells started ringing: *Who the hell was this . . . environmentalist who had snuck into the Bush administration?*

Whitman, oblivious to the firestorm she had created, jumped on the plane to Trieste.

In the days after her appearance on *Crossfire*, a number of high-level executives from coal, utility, and railroad companies (they called themselves the "coal-based generators") visited Washington to warn that Bush's promised CO_2 mandate would "kill coal" and to lobby for government subsidies for clean coal instead. At one meeting with Democratic senator Harry Reid, a strong supporter of greenhouse gas reductions, Irl Engelhardt dismissed global warming as "a PR tool for environmentalists." *Newsweek* reported that Thomas Kuhn, the powerful head of the Edison Electric Institute and Bush's former roommate at Yale, personally called senior White House aides to argue that the CO_2 pledge be dropped. But, Kuhn told the magazine, "it doesn't take any special phone calls from me to make our views known." Myron Ebell, global warming expert at the Competitive Enterprise Institute and one of the most strident voices against CO_2 limits, was not bashful about his views. "This is a colossal mistake," he said of the Bush administration's plans. "If they persist, there will be war."

The good old boys at the Southern Company were as shaken up by Whitman's comments as anyone else in the industry. Southern had as much clout with the new administration as Peabody did, and to help resolve things, it brought in its big gun, lobbyist Haley Barbour. Just as he would do in the coming months in the fight over New Source Review, Barbour made his presence known on the issue of CO_2. On March 1, 2000, he sent out a memo titled "Bush-Cheney Energy Policy and CO_2." The memo was addressed to Cheney, whose energy task force was then meeting every Wednesday, as well as to Energy Secretary Spencer Abraham and other Bush cabinet members. Pointedly omitted from the list were Whitman and her only real ally in the cabinet, Paul O'Neill.

In the memo, Barbour stressed the nation's "serious energy problems," which he said amounted to "an energy crisis" in some regions. Then, in a passage that is a small masterpiece of political rhetoric, Barbour threw down the gauntlet, daring the Bush administration to separate itself from its Democratic predecessors and suggesting that a refusal to do so would align the president with the environmental extremists who wanted to shut down the economy. "A moment of truth is arriving," Barbour wrote. "The question is whether environmental policy still prevails over energy policy with Bush-Cheney, as it did with Clinton-Gore. Demurring on the issue of whether the CO_2 idea is eco-extremism, we must ask, do environmental initiatives, which would greatly exacerbate the energy problems, trump good energy policy, which the country has lacked for eight years?"

It's worth pausing to consider the term "eco-extremism," which Barbour used in his memo and which comes up frequently in debates about global warming. It is one of those political code words used to refer to a mostly mythical group of people who don't understand or deny the link between fossil fuel consumption and civilized life and who, more important, value the health of the planet more than they value the health and prosperity of human beings. Politically, the term is a winner. By deploying the straw man of eco-extremism, coal advocates are able to cast themselves as humanists — as people who are in the business of burning coal because it helps the poor and the vulnerable. Thus, if you take a position that impinges in any way on the free and unfettered consumption of coal, you're an extremist who elevates the birds and the bees above the success of the human race. This idea, or a variation of it, runs through the rhetoric of Big Coal like a thick seam of anthracite. But no one is more frank about it than a scientist named Arthur Robinson, who has likened limiting CO_2 emissions to "technological genocide."

Robinson, head of an organization called the Oregon Institute of Science and Medicine, is best-known as the man behind one of the pillars of global warming skepticism: the so-called Oregon Petition, a 1998 document signed by 17,000 "scientists" which claims that "there is no convincing scientific evidence that human release of . . . greenhouse gasses is causing or will, in the foreseeable future, cause catastrophic heating of the Earth's atmosphere and disruption of the Earth's climate." The petition is widely cited in global warming debates. James Schlesinger, a Peabody Energy board member and former secretary of energy, referred to it in a recent opinion piece in the *Wall Street Journal* denouncing "the theology of global warming." Nebraska senator Chuck Hagel called the 17,000 signatures an "extraordinary response" and cited the petition as his basis for continuing to oppose a global warming treaty. The petition also has been mentioned as evidence of a lack of scientific consensus on global warming in many publications, including *Newsday*, the *Los Angeles Times*, the *Washington Post*, the *Austin American-Statesman*, and the *Denver Post*.

You don't have to dig very deep, however, to discover that Robinson is a unique breed of scientist and that the Oregon Petition was mostly a fraud. Of the 17,000 signers of the petition, only 1,400 claimed to hold a Ph.D. in climate-related science. How many of those signers were legitimate? *Scientific American* took a random sample of 30 of those signatories. Of the 26 the magazine was able to identify, 11 said they still agreed with the petition — 1 was an active climate researcher, 2 had relevant expertise, and 8 signed based on an informal evaluation. Six said they would not sign the petition today, 3 did not remember any such petition, 1 had died, and 5 did not answer repeated messages. "Crudely extrapolating," *Scientific American* reported, "the petition supporters include a core of about 200 climate researchers — a respectable number, though rather a small fraction of the climatological community."

As for Robinson's scientific credentials, the Oregon Institute of Science and Medicine is just another name for a barnlike structure on Robinson's 340-acre retreat in Cave Junction, Oregon, a small town (population 1,126) near the California border. Six people are listed on the board of the institute, "but it's really just me and a few other people who do the work," Robinson explained to me in a telephone conversation. A decade ago, Robinson was a rising star in the world of biochemistry. After getting a Ph.D. from the University of California, he became a protégé and partner of Linus Pauling, the brilliant and controversial chemist and two-time Nobel laureate. In the 1970s, Robinson justifiably challenged Pauling's claims that vitamin C was a cancer cure-all. Pauling called Robinson's work "amateurish" and ended their partnership. (Robinson eventually sued him and won a $575,000 settlement.) Robinson then retreated to Oregon and devoted himself to topics far afield from the scientific mainstream, including work on a molecular clock that he hopes will someday allow scientists to manipulate the aging process. Among Christian fundamentalists, Robinson is best-known for his home-schooling program, which he sells on his Web site for $190. Although he has kept out of the debate over teaching creationism in schools, he told me that he believes evolution is "a hypothesis that is unproven."

Robinson sees the battle over global warming as part of an epic struggle between the rights and faith of the common man and a global elite bent on domination and control. Consider, for example, the views he expressed in a recent essay titled "Science, Politics and Death," which appeared in the *New American,* a publication of the ultraconservative John Birch Society. The essay is full of boilerplate rhetoric of global warming skeptics — the earth's current warming is entirely natural, CO_2 is good for us, and climate models are unreliable. But the essay parts company with more principled skeptics in its portrayal of global warming as a plot cooked up by "environmental extremists" that will cause millions of poor people to "slip backward into the dim twilight in which they suffer silently amid poverty, disease and death." According to Robinson, this poverty, disease, and death are "the true objectives of many environmental extremists who want to radically reduce the world's population, and they are means to an end for a global power elite."

In the final paragraph of the essay, Robinson warns of a coming apocalypse:

For those of us who also understand our accountability to our Creator, the myth of human-caused global warming raises a much larger issue. Thomas Jefferson framed this issue more than two centuries ago when he wrote: "Indeed, I tremble for my country when I reflect that God is just; that His justice cannot sleep forever." Americans today permit the slaughter of millions of children each year by abortion. We are also complicit in the deaths of two to three million children each year in underdeveloped countries by denying them the protection from malaria that DDT provides. Will we now allow an even greater act of technological genocide? Will we participate in a thinly disguised program of world "population reduction" through United Nations constriction of the use of coal, oil and natural gas? Are we to allow the deaths of tens of millions of people from yet another American-supported genocidal action? How much longer do we think God will stay His hand from administering the justice Jefferson feared — unless we do something to stop this pseudo-environmentalist madness?

Not long after I read this essay, I telephoned Robinson to ask if he genuinely believed that reducing our dependence on fossil fuels was a form of technological genocide. Isn't genocide another word for mass murder?

"I'm sorry to have to use terms like that," he said in a very gentlemanly tone. "But I think it's important to speak the truth. When you start cutting back on coal and oil, what you're really talking about is depriving millions of people in places like Africa access to cheap energy to improve their lives. One of these days, people will start to see global warming for what it is — a thinly disguised scam by corporations, the United Nations, and big environmental groups to reduce the world's population. Speaking as a scientist, I can tell you that most people who tout global warming are liars, and the sooner we recognize that, the better."

In Trieste, Whitman was blissfully unaware of the storm brewing back home. She assured her European counterparts that Bush's commitment to CO_2 limits was real and that global warming was a problem that he intended to deal with. Not surprisingly, she was a big hit. The *Financial Times* noted that delegates were "pleasantly surprised" and "impressed by the stance taken by Whitman, head of the U.S. delegation, who calmed fears that the U.S. would ignore the problem of global warming." Featured prominently in the *Times* piece was another unequivocal quote from Whitman: "The president has said global climate change is the greatest environmental challenge that we face and that we must recognize that and take steps to move forward."

Whitman knew that this comment was going to cause trouble back home, and on the return flight to Washington, she wrote a memo to the president. She emphasized that the world community was convinced of the need for immediate action on climate change and that world leaders believed that without U.S. involvement, significant progress would be impossible. She also pointed out that because expectations of the United States were so low, the country was in a position to build goodwill with its allies without endorsing the specifics of Kyoto. She ended the memo with this: "I would strongly recommend that you continue to recognize that global warming is a real and serious issue . . . Mr. President, this is a credibility issue (global warming) for the U.S. in the international community. It is also an issue that is resonating here at home."

At the same time Whitman was writing her note to the president, four Republican senators — Chuck Hagel of Nebraska, Larry Craig of Idaho, Jesse Helms of North Carolina, and Pat Roberts of Kansas — were writing another memo to him. All of them opposed regulating CO_2. The senators' letter, seeking a "clarification of your administration's policy on climate change," quoted some of Whitman's public comments about the urgency of global warming. "We look forward to working with you and your administration on the development of a comprehensive national energy strategy that is environmentally and economically sound, and a common sense, scientifically sound climate change policy," the senators wrote. "However, we need to have a clear understanding of your Administration's position on climate change, in particular the Kyoto Protocol, and regulation of carbon dioxide under the Clean Air Act."

When Whitman returned to Washington a few days later, she was startled to see how things had unraveled while she was away. Whitman and Paul O'Neill agreed that the senators' letter looked suspicious. "The timing, the tone, the emphasis on the senators' desire to work with the administration on a 'comprehensive national energy strategy,' with all environmental issues as a subordinate clause beneath the dictates of energy and economics. That was right out of

Dick Cheney's mouth," observed Ron Suskind in *The Price of Loyalty,* an account of O'Neill's brief tenure in the Bush administration.

"You know what?" O'Neill told Whitman. "I bet they didn't dream up the idea of writing this letter on their own . . . I wouldn't be surprised if you found out that the White House requested this letter of clarification and that the Vice President was preparing the response." O'Neill, who had known and worked with Cheney for years, understood that this was typical of how he operated behind the scenes, manufacturing a situation to get the result he wanted, but without having it be traceable back to him.

Whitman learned that EPA staff had been called to meetings at the White House while she was away and that the administration was preparing to reverse its pledge to regulate CO_2. "Apparently, everyone in those meetings was using the California energy crisis to justify the reversal of the cap," Whitman later observed. Still, she believed that the president would consult her before any decision was made and that she had time to argue her case. A meeting with the president was set for the morning of March 13.

When she arrived at the Oval Office, Whitman took a seat on the edge of the couch, with the president seated in the wing chair near the fireplace. Knowing her time was limited, she launched into a discussion about the indisputable scientific evidence of global warming, as well as the issue of U.S. credibility.

Bush cut her off. "Christie, I've already made my decision," he said. He had already prepared a letter to send back to the senators. He read parts of it to her, mentioning a new Department of Energy report that showed how caps on CO_2 emissions would lead to a dramatic shift away from coal to natural gas for electric power generation and thus to significantly higher electricity prices. With the California energy shortage and the other western states worried about price and availability this summer, he told her, we just can't harm consumers.

Whitman was dumbfounded. She knew as well as anyone that this "new" Department of Energy report was flawed and, more important, that the idea that doing anything about global warming would inevitably lead to significantly higher electricity prices was, at the very least, debatable. Whitman also knew about four other reports on this issue that had been prepared in the past six months. For example, a study by the Department of Energy in November 2000 had found that CO_2 emissions could be reduced to 1990 levels with a net increase in Americans' energy bills of less than 1 percent in the year 2010.

But the president didn't want to hear that. Nor did he want to hear any suggestion that rolling blackouts in California — which, as it turned out, were caused more by corrupt and greedy energy traders than by a supply shortage — were not an excuse to deny the urgency of addressing global warming. He also wasn't interested in hearing about the political costs of alienating America from its international allies on this issue.

All in all, it was a stunning move. Whitman knew, in that instant, not only that the administration would do nothing about global warming, but also that Bush was a different kind of man than she had taken him to be. He had allowed her to run around the world, putting her own credibility on the line and using her hard-won political respectability to elaborate on a commitment that he had made to the American public. When his pals in the coal and oil business had kicked up a fuss, he had broken that commitment without even discussing it with her.

Whitman might have offered the president her resignation right there, but instead she hung around for two more years, taking bullets for the Bush administration on the New Source Review rollbacks and mercury regulations, enduring jokes from cabinet members (Secretary of State Colin Powell called her the administration's "wind dummy" — a military term for the object you throw out of an airplane to see which way the wind is blowing over a landing zone), and looking for a graceful exit. By May 2003, she had had enough. She resigned, saying that she wanted to spend more time with her husband at their New Jersey farm. When she broke the news to the president, he gave her a hug.

Irl Engelhardt fared considerably better. The reversal of Bush's campaign pledge removed the most significant obstacle to the coal industry's future, and to Peabody's. This was underscored two months later when Vice President Cheney's energy task force published its final report. Although the report made the obligatory nods to renewable energy and energy conservation and efficiency, it was really not much more than a drill-and-burn manifesto for the fossil fuel industry, including Big Coal.

Peabody took full advantage of it. Exactly five days after the energy plan was released, Peabody's initial public offering occurred on the New York Stock Exchange under the ticker symbol BTU (an abbreviation for "British thermal unit," the standard measurement of a unit of energy). The company offered 15 million shares, whose price soared more than 30 percent during the first day of trading, closing at $36 a share and raising more than $430 million for the company — $99 million more than expected. Peabody was now not just the biggest coal company in the world; it was also one of the richest. And Engelhardt did pretty well, too. According to SEC filings, as of May 31, 2001, he owned, directly and indirectly, more than 600,000 shares of Peabody stock. He held on to the bulk of those shares until 2005, when the stock price hit $86. Then he decided it was time to go. He announced that he would step down as CEO of Peabody as of January 1, 2006, and began cashing out. By the end of 2005, he pocketed more than $55 million. Not bad for a poor kid from Pinckneyville, Illinois.

Chapter 9

The Coal Rush

OH, GLORY DAYS! America the beautiful! Opportunity returns! The governor is coming to Nashville! On a cold morning in February 2005, the gym at Nashville Community High School in southern Illinois was jammed to the rafters with kids. More than 2,300 students had been dismissed from their morning classes and bused in from all around the region. Squads of cheerleaders cartwheeled across the gym floor, while the Nashville Hornets school band filled the gym with rah-rah songs. On the walls behind a hastily erected podium, ready for the cameras, was a banner that read OPPORTU-NITY RETURNS, Governor Rod Blagojevich's campaign slogan to bring prosperity back to the coalfields of southern Illinois.

Soon the band stopped playing and the gym fell into a hush. Nashville mayor Raymond Kolweier introduced the first Democratic governor of Illinois in more than twenty years, a man whom syndicated columnist George Will had already singled out as a possible presidential candidate. But first, the forty-four-year-old governor had to win reelection in 2006, and to do that, it was vital that he convince the people of southern Illinois that he was not just an Elvis-loving dweeb from the north. The best way to accomplish that was to get behind coal.

So Governor Blagojevich stepped to the podium and quickly got down to the business of touting the $2 billion coal-fired power plant that Peabody Energy was preparing to build just a few miles south of Nashville. Peabody's intention to build the plant, which was officially called the Prairie State Energy Campus, had been announced not long after the company went public in 2001. Now, after several years of controversy and rumor, Prairie State was on the verge of becoming a reality. The state had granted the final air permit a few days earlier, and the Illinois Finance Authority had agreed to offer unspecified millions in low-interest bonds to help finance the plant. Prairie State, the governor boasted, would create 2,500 construction jobs, 450 permanent jobs, and $100 million or so a year in spin-off revenues for the region. "Illinois coal is experiencing a rebirth," Governor Blagojevich told the crowd, "and I can think of no better example of that rebirth than the Prairie State Energy Campus."

A phalanx of Peabody executives was on hand to show its support of the governor's vision. Irl Engelhardt, dressed in a dark suit and tie, stepped up to the microphone and told a rambling joke that compared Peabody's efforts to build Prairie State to Noah's efforts to build his ark. Engelhardt warned that outsiders from far away, who were part of "an international agenda," were trying to stop the construction of the plant, without justification. "The technology Prairie State will use is absolutely the best that has been put together on a coal plant," Engelhardt reassured the students. "Prairie State is an important step forward in terms of the cleanliness of coal plants, and ultimately will help us get to near-zero emissions from coal plants."

Whether Prairie State can accurately be described as an important step forward depends a lot on where you're standing. Technology-wise, Prairie State is pretty much the same old beast that Big Coal has been building for years, tricked out with a more efficient boiler and the latest scrubbers. Emissions of sulfur dioxide and nitrogen oxides will be far below those from the old coal burners but still nowhere near as low as those from a natural gas plant. But the real problem with Prairie State is CO_2. Prairie State will emit more than 11 million tons a year, marginally less than a similar size coal plan built thirty years ago, but more than twice as much as every vehicle sold by the Ford Motor Company in a single year. In a world that took global warming seriously, building an old-fashioned plant like Prairie State might be considered an important step backward. But that is not the world we live in.

Governor Blagojevich, of all people, knew the score here. He prided himself as a "third way" politician, not beholden to special interests or partisan orthodoxy. In the months after he took office in 2002, you could almost make out the outlines of a politician who had, as he would later put it, "the testicular virility" not only to reinvent Illinois's rust belt economy but also to banish the ghost of Samuel Insull from the very state where he built his empire. Illinois, after all, has lots of wind for wind turbines, lots of crops for biofuels, and lots of bright minds trying to figure out ways to break the old paradigms and rethink America's energy problems. It's also the home of the Chicago Climate Exchange, an innovative carbon trading marketplace that will play an important role in helping big coal burners manage their CO_2 allowances if and when the United States decides to limit greenhouse gas emissions.

Unfortunately, after a few years in office, the governor's "third way" was starting to look a lot like the old way. He had done a number of laudable things, such as requiring power companies to buy 8 percent of their electricity from renewable sources by 2013 and calling for tougher restrictions on mercury emissions from Illinois coal plants. But then he'd turned around and lent his support to the construction of a coal plant like Prairie State, which will pump more CO_2 into the atmosphere in one year than a fleet of wind turbines will save in fifty years. Supporting the plant with Illinois tax dollars — especially when it is being built by the biggest, richest coal company in America — was about as forward-thinking as subsidizing a fleet of Hummers for Chicago meter maids.

But lest anyone think the governor was merely pandering for votes in southern Illinois, Blagojevich spoke about the parallels between the steel mills, where his father, a Serbian immigrant, worked for years, and the Illinois coal industry. He also spoke about the higher purpose of this project, reminding the students that America was at war in the oil-rich Middle East. Supporting the coal industry was not just good for Illinois, he implied, it was good for America.

A few local politicians chimed in, the band struck up the Hornets' fight song, and there was a lot of clapping and backslapping. Even the kids up in the bleachers, most of whom had been born long after the coal industry had died in this region, were standing on their feet, cheering the governor who was going to bring back the days when blue-collar jobs were plentiful, the economy was booming, and Marcus Welby was everyone's family doctor. It was an appealing vision — a mix of optimism, nostalgia, and good old American can-do spirit.

"Coal is USA!" someone up in the bleachers shouted. "Coal is USA!"

For Big Coal, the slogan "Opportunity Returns" is, if anything, an understatement. According to the International Energy Agency (IEA), the equivalent of fourteen hundred 1,000-megawatt coal-fired power plants will be built in the world by 2030. About half of those plants will be in China, 15 percent in India, and the remaining third primarily in the coal-burning West, including Australia and the United States. In America, the IEA predicts, about one third of the new electric capacity built between 2005 and 2025 will be coal fired. Besides Peabody's Prairie State plant in Illinois, as of 2006 there were more than one hundred and fifty new coal plants proposed in more than thirty states, including Florida, Wisconsin, Montana, Washington, New York, New Mexico, North Dakota, and Texas.

In addition, as oil prices climbed above sixty dollars a barrel in 2005, there was renewed interest in building new coal-to-liquids plants that can transform coal into synthetic diesel and other fuels. There is nothing new about this technology — Adolf Hitler used it to make fuel for his tanks and fighter planes during World War II — but it is expensive, inefficient, and CO_2-intensive. Even so, as fear that the world's remaining oil reserves may be depleted increases, so does desperation to find ways to replace oil. China is investing $24 billion in a number of coal-to-liquids plants, and proposals for new plants have been floated in at least seven U.S. states, including West Virginia and Montana. There is no question that converting coal into synthetic fuels can be done, and, if oil prices stay high, it can even be done at a profit. But does that mean it's a good idea?

Big Coal frequently argues that today's coal boom is not like coal booms of the past. "This is not your grandfather's coal industry," Bret Clayton, CEO of Kennecott Energy, told me. And in some ways, that's true. Mining practices in some regions of the country have improved, and the industry is safer than it was thirty years ago. Emissions of sulfur dioxide and nitrogen dioxide at new coal plants are much lower than they used to be. "Increasingly clean" is the industry's favorite sound bite.

But the word "increasingly" carries a heavy load for Big Coal. The scrubbers on new coal plants might be better, but new plants still release massive quantities of pollution into the air, require prodigious amounts of water for cooling, and generate millions of tons of heavy metal–laden coal ash. In addition, "increasingly clean" does not exactly describe the mountains of Appalachia that are being blasted away to supply the fuel for these new plants. Most important, this new generation of coal plants is very much like the previous generation in one significant way: they pump hundreds of millions of tons of CO_2 into the atmosphere. By doing so, they virtually eliminate any chance we have of reducing the impact of global warming to a manageable level.

"To put it plainly," argues Tom Burke, former director of the environmental organization Friends of the Earth and now a consultant to Rio Tinto, the world's largest mining company, "if these new coal plants are built at anywhere near the rate that the IEA projects, the odds that our grandchildren will live in a world with a stable climate are about zero."

The calculation is fairly straightforward. According to a number of prominent scientists and climate modelers, the world as we know it can tolerate about 3.5 degrees Fahrenheit of warming (about 2 degrees Celsius) without major disruptions. This doesn't mean there won't be death, disease, and hardship, but in general, there will be pluses and minuses of the warming trend. Long droughts will be offset by increased growing seasons in northern latitudes, for example. In addition, the risks of catastrophic climate change are estimated to be relatively small.

Above 3.5 degrees Fahrenheit, however, all bets are off. At that point, climate models begin to show more severe disruptions of precipitation patterns, epic droughts, increasingly intense storms and hurricanes, and rapidly dying coral reefs. The risk of ocean conveyor disruption rises, as does the threat of other potentially catastrophic events, such as the collapse of the Amazon rain forest, which would have a devastating impact on the earth's ability to cycle carbon.

How much leeway does that leave us right now? As of 2005, the earth has already warmed about 1 degree Fahrenheit (.6 degree Celsius) since preindustrial times. It is generally agreed that there is about 1 degree Fahrenheit of warming built into the climate system already, mostly due to the heat that is stored in the oceans, which will gradually be radiated back into the atmosphere in the coming years. So for the purposes of this calculation, we're already at 2 degrees Fahrenheit of warming. Over the past decade, the level of CO_2 in the atmosphere has been rising at a rate of about 1.5 parts per million each year. To limit warming at 3.5 degrees Fahrenheit, climate researchers believe we will have to stabilize CO_2 levels at between 400 and 500 parts per million. Which means that, at the rate we're going, we'll cross into the red zone around the year 2017. "Climate change is real," warned the usually circumspect Rajendra Pachauri, head of the United Nations Intergovernmental Panel on Climate Change, in early 2005. "We have just a small window of opportunity and it is closing rather rapidly. There is not a moment to lose."

This is why the coal boom is so alarming. Right now about one quarter of the world's CO_2 emissions come from coal. If we go ahead with these new coal plants, they will add roughly 570 billion tons of CO_2 to the atmosphere over the life of the plants. (To put that number in perspective, 570 billion tons is about as much CO_2 as released by all the coal burned in the past 250 years.) If that happens, our chances of stabilizing the climate are virtually zero. According to Malte Meinshausen, a widely respected environmental physicist at the Swiss Federal Institute of Technology in Zurich, to have a likely chance of limiting warming to 3.5 degrees Fahrenheit, CO_2 emissions, instead of rising by 100 percent or more by 2050, need to fall by 10 to 50 percent. And there is no time to waste. Meinshausen calculates that delaying emissions cuts by ten years would nearly double the required reduction rate in 2025.

This is the bottom line: even if every SUV were downsized to a

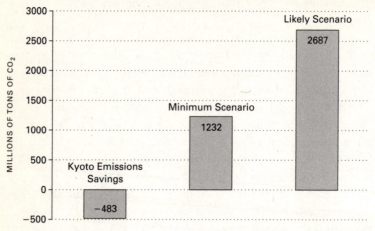

NEW COAL PLANTS SWAMP KYOTO

By 2012, expected cuts in greenhouse gas emissions under the Kyoto treaty will be swamped by emissions from a surge of new coal-fired plants built in China, India, and the United States.

Source: *Christian Science Monitor,* Dec. 27, 2004

Schwinn, every truck and bus repowered to burn biodiesel, and every refrigerator retrofitted to run with solar panels, if we keep burning coal the old-fashioned way, we are playing Russian roulette with the very thing that makes our life on earth possible — a steady, temperate climate.

When Peabody Energy announced its plans for the Prairie State Energy Campus on October 22, 2001, the future was very much a part of the plant's rationale — but it was not the future of the planet, it was the future of Peabody's stock price. One big problem for coal mining companies has always been growth. Supplying coal to the four hundred or so big coal plants around the country might give a company like Peabody a growth rate of 1 or 2 percent a year, but that is not enough to keep Wall Street happy. This is a problem for every coal company, but it was a particular problem for Peabody in 2001. Its Powder River basin operations were going like gangbusters, but the company also owned huge reserves of Illinois coal — a legacy that goes back to the days when the company supplied coal for Samuel Insull's power plants in the Midwest — which nobody was interested in because it was expensive to mine and loaded with sulfur.

What to do? If you can't find a power plant that wants to burn your coal, one way to solve that problem is to build one yourself — which is exactly what Peabody proposed with Prairie State. The plant would not only be designed to burn Peabody coal, but it would also be built right beside the coal mine — a "mine-mouth plant." "Most power plants are built in order to generate electricity," Bill Hoback, head of the Illinois Office of Coal Development, explained to me. "Prairie State was really conceived more as a platform to burn Peabody coal."

In the old days of the electric power business, before markets began to be restructured and were open to competition, utilities had to conduct detailed studies of future electric power demand to justify the construction of new power plants to regulators. But in restructured markets like Illinois, none of that is necessary anymore. As Michael Skelly, head of development at Horizon Wind Energy, one of the largest wind companies in America, puts it, "The only question you have to answer now is, can you build something that will generate electricity cheaper than your competitors?"

To be sure, there are other factors beyond price that influence what kind of power plant gets built where. No one will erect wind turbines where there is no wind, solar panels where there is little sun, or a coal plant where there is no easy access to coal. Access to transmission lines is also a major factor, as are hard-to-quantify factors such as how difficult the permitting process will be. In places such as Illinois, where the coal industry has taken a beating in recent years and there are many out-of-work coal miners, companies such as Peabody can expect a lot of help from local politicians to cut through red tape.

But beyond that, it really comes down to numbers. To put it simply, coal plants are popping up everywhere in America right now because they are the cheapest way to generate electricity. Unless there is a revolution in the way electricity is priced and marketed (highly unlikely), all the hidden costs of burning coal that are offloaded to the public — the blighted mountains and economic ruin of many coal mining regions, the medical bills related to the heart attacks and asthma caused by air pollution, the lost future incomes of kids adversely affected because their mothers ate mercury-laden fish when they were pregnant — are unlikely to be calculated into the price of power anytime soon.

The X factor in electricity pricing, however, is global warming. How do you calculate into the price of power the risks associated with abrupt climate change? This is by far the biggest looming threat to the hegemony of cheap coal. Despite the industry's denials about the reality of climate change, almost everyone acknowledges that in the next decade or so, laws that put a cap or quota on CO_2 emissions will be implemented in America. And when that happens, when electric power companies have to pay a financial price for the right to release millions of tons of CO_2 into the atmosphere, coal's position as a cheap power source will be seriously threatened. In fact, in a perverse way, it is precisely this threat of future regulation that is fueling the mad dash to throw up coal plants all over the country. If the plants aren't approved and built now, the reasoning goes, once a price is attached to CO_2 emissions, costs will go up, and coal plants will find themselves being underbid by wind turbines, gas plants, and other forms of generation that have less carbon liability. What's going on now is not exactly a land grab — it's an atmosphere grab.

For Big Coal, there is another option: instead of throwing up the same old carbon-intensive coal burners, they could move to a more flexible power plant technology that not only is more efficient but also allows for the possibility of someday capturing and sequestering the CO_2 from the plants. There is just such a new technology waiting in the wings. It goes by the unfortunately complicated name of integrated gasification combined cycle, or IGCC.

The difference between an IGCC power plant and a conventional coal plant is sometimes said to be akin to the difference between a Toyota Prius and a Chevrolet Suburban. But it's really more fundamental than that. Instead of burning coal in a big steel box like conventional coal plants do, IGCC plants use heat and pressure to cook off the impurities in coal and convert it into a synthetic gas; the gas is then captured and burned in a turbine. The advantages of IGCC plants are many. They are 10 percent more efficient than conventional coal plants, consume 40 percent less water, produce half as much ash and solid waste, and are nearly as clean burning as natural gas plants. But more important, it is far easier and cheaper to capture CO_2 from coal at an IGCC plant than at a conventional coal plant. In theory, the CO_2 that is removed from the coal can be stored in underground aquifers, or perhaps under the sea. Geologic carbon storage is still a controversial and largely untested idea, and one that would be more economical in some regions of the country than in others. But the combination of IGCC with CO_2 capture and storage at least offers a plausible way to continue using coal without willfully trashing the climate. The National Commission on Energy Policy — an independent, bipartisan advisory body — concluded in a 2004 report that "the future of coal and the success of greenhouse gas mitigation policies may well hinge to a large extent on whether this technology can be successfully commercialized and deployed over the next 20 years."

Politicians in coal-rich states such as Illinois love IGCC. For one thing, it offers a way to create mining jobs without whipping coal-hating environmentalists into a frenzy. For another, gasification is of tremendous interest to farmers, who have been hit hard by the rising price of fertilizers made from natural gas. One of the many virtues of gasification is its flexibility. The gas can be burned in a turbine to create electricity, but its elements also can be broken down to create the building blocks for fertilizers and other useful chemicals. Anytime you can get farmers, environmentalists, and power companies marching in the same direction, that is very good news for an election-minded politician.

For Illinois, pushing IGCC could help the state shrug off its rust belt image and promote itself as a leader in green innovation. Chicago mayor Richard Daley, for example, is often singled out as one of the most environmentally progressive mayors in the country. He has built bike paths, pushed renewable energy, upgraded building codes for better energy efficiency, and promoted a "green roofs" program to get people to plant trees and gardens on urban rooftops. He also has been an outspoken critic of President Bush's rejection of the Kyoto Protocol, as well as his general failure to speak frankly about the consequences of global warming. (Illinois's two senators, Barack Obama and Dick Durbin, also are critics of the president on this point.) Given that it is politically and economically impossible for any Illinois elected official to abandon coal, pushing hard for coal gasification could help suggest that whatever the state's problems might be, denial of one of the biggest environmental and economic issues of our time isn't one of them.

Unfortunately, what's good for the climate and for the state of Illinois is not necessarily what's good for Peabody Energy. What does a company like Peabody get for building an IGCC plant? In a deregulated market, the benefits all accrue to the public, not to Peabody's shareholders. Peabody is in the business of making money and expanding its dominion, not saving the planet. As long as it was still legal to throw up an old coal burner, that was exactly what the company was going to do.

Of course, Peabody executives didn't put it that way in public. When asked why they didn't propose an IGCC plant at the Prairie State site instead of a pulverized coal plant, they mostly talked about how the coal in the area had too much sulfur for gasification (a view that most coal gasification experts I talked to disagreed with), how IGCC plants weren't proven to be reliable, and, most of all, how IGCC plants were still 20 percent more expensive than pulverized coal plants. "While we support IGCC," Irl Engelhardt said shortly after the rally in Nashville, "America needs to build generating plants now. And the ones we build need to work. It's not prudent to deploy technology until it is reliable."

Big Coal is much better at touting new technology than actually putting it to work. There is no better example of this than FutureGen, the Department of Energy's $1 billion zero-emissions plant that will turn coal into electricity and hydrogen. FutureGen was proposed by President Bush in 2003 and has been backed by a consortium of coal and electric power companies, including Peabody Energy and the Southern Company. The goal is to figure out a way to make dirty coal truly "clean," as well as use America's vast reserves of coal to create hydrogen, the much-hyped fuel of the future.

It's just as likely, however, that FutureGen — or "NeverGen," as it's affectionately known to some people in the industry — will turn out to be just another expensive government boondoggle. The 275-megawatt prototype plant is scheduled to be operational by 2012. Even if it is up and running by then (which many people in the industry consider a highly dubious prospect), it will take several decades for engineers to tweak the design and build spin-off plants. Any number of interesting engineering breakthroughs could come out of FutureGen, and as basic research it may well be a worthwhile project. But it would be foolish to bet on FutureGen as a solution to America's energy problems. More important, by the time the plants are perfected and deployed, the global warming genie is likely to be out of the bottle.

Pessimism about FutureGen was apparently shared by many members of Congress in 2004, when they voted to cut funding for the project to a minimum. Although funding was restored in 2005, it's hard to find anyone without a vested interest in the project who believes that FutureGen is anything but an expensive political decoy to make it look like the coal industry is doing something big and important to address global warming, while in fact it is doing very little. "The coal industry is very good at touting technology that is twenty years away forever," says Dale Simbeck, a respected power industry consultant who specializes in gasification technology.

The truth is, if power companies wanted to build a near-zero-emissions coal plant, they could do it today. IGCC, combined with scrubbers and CO_2 capture and storage, would reduce emissions close to zero. But of the one hundred and fifty or so coal plants that are being planned or are under construction in America right now, only a handful use IGCC technology.

Big Coal has used two main arguments to explain why it has failed to embrace this technology, both of which are typically short-sighted. The first is economic. As of 2005, on a straight cost basis, an IGCC plant is 10 to 20 percent more expensive to build than a conventional coal plant. In regulated states such as Wisconsin, utilities have successfully persuaded regulators that this price hike is unjustified. Why pay more for electricity if you don't have to?

But this is a narrow view of the situation. In fact, the few companies that have decided to plunge ahead have come to precisely the opposite conclusion. As Mike Mudd, head of new generation at American Electric Power (AEP), explained to the Ohio Public Utilities Commission in 2005, if you factor in the obvious reality of global warming and the equally obvious political reality that CO_2 limits are coming in the next decade or so, it's the old-fashioned combustion coal plants that don't make long-term economic sense today. According to Mudd's analysis, which takes into consideration projected fuel prices and other operating costs of a power plant coming online in 2010, building and operating an IGCC plant is less than 10 percent more expensive than building a pulverized coal plant: $56 per megawatt-hour for IGCC vs. $52 per megawatt-hour for pulverized coal. But as soon as you add in the advantages of carbon capture and storage, the numbers favor IGCC by a wide margin: $79 per megawatt-hour for IGCC vs. $95 per megawatt-hour for pulverized coal. Mudd's conclusion: "It's AEP's opinion that taking this first step into the new era of electric generation using [gasification] technology is both fiscally responsible and the right thing to do as a matter of public policy, both for AEP and for Ohio."

The second argument is that IGCC is still an unproven technology. This claim is true only by the most conservative standards. Engineers have been gasifying coal for 150 years (coal gas, also known as "town gas," lit most cities before the arrival of electric power), and gasification is widely used for many purposes in the chemical industry. The only part that's new is combining coal gasification with electric power generation, and although the engineering is complex, it's not alchemy. In the United States, the first coal IGCC demonstration project, the Cool Water plan in California, ran successfully from 1984 to 1989. In the 1990s, coal IGCC plants were built in Indiana, Florida, the Netherlands, and Spain. Ten more IGCC plants that use other fuel stocks — mostly waste products from oil refineries — have been built worldwide. All these plants are still in operation today and represent a combined investment of $8 billion. Some of the biggest names in power plant engineering — GE, Shell, Bechtel — are starting to promote IGCC, but Big Coal has been slow on the uptake. In the short term, the industry would rather dispute the science proving the urgency of global warming and focus on the promise of whiz-bang technology that's forever twenty years down the road. Fighting for the status quo is, of course, nothing new in big business: IBM dissed PCs, the recording industry fought music downloading on the Internet, and General Motors' stubborn attachment to SUVs cost the company billions. But with coal, the stakes are much higher, both for the earth's climate and for the future of the coal industry itself.

When all is said and done, however, the real problem with IGCC is not economic or technological; it is cultural. The "boiler boys," as power industry traditionalists are known, have been burning rocks for 150 years. To them, a coal plant is as elemental as a campfire, and nearly as lovely. IGCC is a chemical process, and there is nothing romantic about an oil refinery. Making the shift from one technology to the other is a huge leap, especially for a conservative, change-averse industry such as coal. Perhaps more significant is the fact that IGCC forces Big Coal to shed its view of the world as one big mountain to be strip-mined and begin thinking long-term rather than short-term, about sustainability rather than exploitation.

Rather than embrace this change, many companies have decided to push for one more generation of old combustion plants, rebranding them as "clean." Given the urgency of the need to address rising CO_2 levels in the atmosphere, this is, at best, misleading. In fact, the rush to throw up an old coal plant, bolt on some new scrubbers, and sell it as "clean coal technology" is a transparently cynical attempt to cash in before the economic hurricane of global warming hits. The executives who are building these plants know that once they are constructed, no one is going to tear them down, giving the companies fifty years of steady profits.

The real game here is one of brinkmanship. The more urgent the problem of global warming becomes, the more pressure politicians will feel to push, prod, and bribe Big Coal to adopt IGCC. And when that time comes, coal industry executives will be standing there with their hands out: *If you want us to change the way we do business, we need help.* The Energy Policy Act of 2005 included $5 billion in subsidies to Big Coal, including $2 billion for the Clean Coal Power Initiative, a part of which will go toward supporting IGCC.

Clean coal subsidies have been widely criticized by taxpayer groups, as well as Congress's own watchdog, the Government Accountability Office, as being wasteful and poorly administered. It's also true that one of the most successful coal gasification plants in the country — the Eastman Chemical gasification plant in Kingsport, Tennessee — was built entirely without state or federal subsidies. The plant recently appeared on the cover of *Power* magazine, the *Vanity Fair* of the power industry, under the blurb "Coal Gasification: Ready for Prime Time." *Power* singled out the Eastman plant, which transforms high-sulfur Appalachian coal into methanol, a chemical used in plastic production, in part because its performance record gives lie to the idea that coal gasification is too untrustworthy for commercial use. Since the plant was built in 1983, it's been extraordinarily reliable, with an outage rate that rivals that of the best combustion power plants. Not only that, but according to Eastman, the plant turns a profit.

But if there is one place where federal assistance may make sense, it's in speeding up the adaptation of IGCC as the industry standard for new coal plants. Unlike combustion coal plants, IGCC is a technology that actually has a future. It just needs a little help getting there. A well-designed, rigorously applied federal program to accelerate the uptake of IGCC, especially combined with a ruthless cutting of pork for combustion coal plants, would have enormous economic and public health benefits. The loan guarantees and other assistance in the Energy Policy Act of 2005 were a start, but much more could be done.

For environmentalists, IGCC poses a dilemma. For the greens who would like to drive a stake through the heart of the coal industry, the efficiency and pollution improvements offered by IGCC are too little too late. Many environmentalists fear that supporting IGCC ensures another fifty years of dependence on fossil fuels, when we really should be focusing all our time and energy on figuring out a way to go cold turkey. "Coal is a nineteenth-century fuel," says Robert Ukeiley, a Sierra Club attorney who battled a new Peabody coal plant in Kentucky. "The sooner we admit that, the better." Even big-picture energy thinkers such as Jesse Ausubel at Rockefeller University in New York believe that IGCC is a last gasp of an industry that is destined for obsolescence. Ausubel compares IGCC to last-minute innovations in clipper ships as they attempted to compete with the new steam-driven ships. "It's hard to fight the larger forces of progress," he says. "IGCC doesn't change the fact that coal is caught on the wrong side of a much larger trend toward the decarbonization of energy."

But to David Hawkins, head of the climate program at the Natural Resources Defense Council and elder statesman of the environmental movement, IGCC is the centerpiece of "a grand bargain" between power companies and environmentalists: you (Big Coal) embrace IGCC, as well as future caps on CO_2 emissions and meaningful reforms to curb the abuse of coal mining regulations, and we (environmentalists) will join you in the fight to get these plants built, including calling for more federal assistance and streamlining the permitting process. So far, Big Coal hasn't shown much interest. In fact, when I asked Peabody's spokesperson Vic Svec whom he considered the "extremists" in the environmental movement, he singled out Hawkins. "I think if you look back over his past statements," Svec told me, "it's fair to say that Hawkins's real agenda is not to promote IGCC, but to shut down the coal industry."

With statements like these, it's clear there's no "grand bargain" in the offing anytime soon — at least not with Peabody. Economists have a phrase for what happens when the public good is allowed to be exploited for private profit: "the tragedy of the commons." It refers to a well-known parable about an English village that shares a single grazing area for livestock. It works well enough, as long as each farmer takes care not to add too many sheep and overgraze the commons. The problems start if farmers get greedy and add more livestock to the commons, taking advantage of the public good for their own benefit. In a few years, the soil is depleted by overgrazing, the commons becomes unusable, and the village perishes.

By building Prairie State, Peabody is the industrial equivalent of the farmer who lets his sheep out into the pasture late at night, hoping to fatten them up for market and therefore sell them for a higher price than his neighbor gets for his sheep. Irl Engelhardt and other Peabody execs tout jobs and the spin-off benefits to the local economy, but those jobs and benefits would be just as big if they built something less exploitative of the commons (in this case, the earth's atmosphere). But they don't, because the real purpose of a coal burner like Prairie State is not to spiff up the village, but to fill the pockets of the renegade farmer.

In the short term, this strategy may well make shareholders happy. In 2004 and 2005, coal stocks were hotter than a boiler's steam tube. In some cases, the windfalls have been downright obscene. In the third quarter of 2006, TXU, the big Texas power company that is proposing to build a fleet of old-style coal burners, recorded a profit of over a billion dollars. But what about in the long term? "When people in the U.S. finally wake up to the reality of global warming, they may react in a very emotional way," says Rio Tinto's Tom Burke, who is based in London. "They may decide, 'Hell, let's slap a fifty-dollar-a-ton tax on carbon.' Or let's just shut the coal plants down. Forget the climate. From a purely financial point of view, you have to be mad to build a conventional coal plant these days. This is going to start dawning on people soon — especially shareholders. That's why I think parts of the U.S. coal industry have to be among the dumbest businesses in the world. They think global warming is just another environmental problem to deal with. They've convinced their favorite politicians of it, too. They really have no idea what's at stake here — either for their companies, for the coal industry, or for the planet."

Will we soon be filling up our gas tanks with coal-derived fuels? Big Coal would certainly like to think so. Although the gasification technology behind coal-to-liquids plants (or coal refineries, as they are sometimes called) is similar to the technology used with IGCC, you don't hear the people who are proposing these plants complaining about how unreliable or expensive it is. On the contrary, they often make it sound as easy as making lemon juice from lemons. Why? Because unlike IGCC, where the benefits go first to the public (cleaner air, lower greenhouse gas emissions), the benefits of coal refineries go right to the bottom line. The math is simple: if oil prices hold steady above sixty dollars a barrel, a coal refinery churning out diesel fuel for forty dollars a barrel is going to make somebody very rich.

Coal liquefaction was pioneered in Germany in the 1920s and

1930s. At the time, experts scoffed at Hitler's idea that he could conquer the world, largely because Germany had almost no domestic supplies of petroleum. Hitler, however, had begun assembling a large industrial complex to manufacture synthetic petroleum from Germany's abundant coal supplies. The Nazis used basically the same process that was commonly used to make "town gas" but added a secondary process, discovered by two German researchers, Franz Fischer and Hans Tropsch, that hydrogenated the coal gas, creating a petroleum substitute. It was a complex, laborious process, but it worked well enough to fuel Hitler's dream of world domination. More than 90 percent of Germany's aviation gasoline and half of its total petroleum during World War II came from synthetic fuel plants. At its peak in early 1944, the German synfuels effort produced more than 124,000 barrels per day at twenty-five plants. In an effort to cripple the Nazi war machine, Allied bombers began targeting synfuel plants, leaving Hitler's planes and tanks high and dry, and helping to bring an end to the Third Reich.

In the 1970s, when America was worried that it might find itself, like Nazi Germany, with more coal than oil, the newly created Department of Energy started pushing the production of synfuels. A number of big-budget research programs were undertaken, but these were more or less abandoned when oil prices suddenly dropped and we went back to our oil-guzzling ways. Congress also passed special tax breaks to encourage the use of "alternative" fuels. Today, these tax breaks are a bonanza for companies like Progress Energy in Raleigh, North Carolina, and DTE Energy in Detroit, Michigan, which have claimed hundreds of millions of dollars in tax credit in recent years by spraying coal with latex or pine-tar resin, magically transforming it into a synfuel (critics call this practice "spray and pray"). In most cases, these synfuels don't burn any cleaner or more efficiently than untreated coal, but to qualify for the tax breaks, the law simply requires that the coal undergo a "significant chemical change." "The entire synfuels industry is basically a sham," says Congressman Lloyd Doggett, a Texas Democrat who has lobbied tirelessly to close the tax loophole. He estimates that synfuel tax breaks cost taxpayers more than a billion dollars each year.

Today's synfuel boosters claim to be a different breed from the hucksters of the past. One of the industry's biggest cheerleaders is Montana's gun-loving, dog-owning Democratic governor, Brian Schweitzer. Montana has large reserves of coal (even more than neighboring Wyoming), but nobody wants it right now because a lot of it is dirty, low-grade coal, and because there are no railroads in the area to haul it out. To Schweitzer, building a new generation of coal-to-liquids plants in Montana would be a good way to put the state's coal to work (because a coal liquefaction plant is built at the mine, railroads aren't an issue, and the technology used in the plant works well with low-grade coal) and bring billions of dollars of economic development money to the state. Schweitzer has proposed building a coal-to-liquids plant capable of producing 150,000 barrels of diesel a day — a modest-size plant, as refineries go — at a cost of some $7.2 billion. To make the proposal easier to swallow, he suggests constructing it in 22,000-barrel-per-day modules, at a cost of about $1.2 billion each.

Schweitzer's pitch is an artful mix of patriotism and progressivism. In an op-ed piece that appeared in the *New York Times,* he wrote, "Like all Americans, Montanans are . . . tired of paying $3 a gallon for gas, tired of watching third-world nations overtake us in energy innovation, and tired of supporting the kind of tyrants that young Americans have spent two centuries fighting and dying to defeat." Synfuels, Schweitzer believes, are a neglected resource, one that can both take the edge off of high oil prices and help America gain energy independence. To his credit, Schweitzer's pitch for synfuels is often accompanied by equally strong words in favor of renewable energy and energy conservation.

As a way of creating backup fuel for the military, coal-to-liquids plants make sense. A few billion tax dollars spent to ensure that the air force will have plentiful fuel for its fighters no matter what happens to Middle Eastern oil reserves is a prudent investment. But coal-to-liquids plants are not the answer to America's energy problems. For one thing, they make diesel, not gasoline. For a variety of technical reasons, gasoline is much more difficult to brew from coal and is beyond the capability of today's technology. For another, coal-to-liquids plants are so complex and expensive that it's difficult to imagine a scenario in which enough plants could be built quickly enough to make more than a small dent in America's 20-million-barrels-a-day oil habit. Finally, these plants consume prodigious quantities of water, both as part of the coal conversion process and for cooling. According to Andre Steynberg, the technical manager for research and development at Sasol, a South African firm that builds synfuel plants, a typical plant consumes about three and a half barrels of water for each barrel of synthetic fuel produced from coal. In many parts of the west, where water is an increasingly precious commodity, that is not a good bargain.

But the biggest problem is that coal-to-liquids plants are a step backward on global warming. They siphon money and attention away from other alternatives to imported oil, such as cellulosic ethanol (a fuel made from forest and agricultural wastes such as cornstalks and wood chips, as well as "energy crops" such as switchgrass) that can reduce greenhouse gas emissions by as much as 85 percent compared to conventional gasoline. In contrast, the amount of CO_2 released by refining coal can be 50 to 100 percent higher than refining petroleum, depending on the technology that is used. In theory, the CO_2 from these plants could be captured and sequestered underground, similar to the way it could be with an IGCC plant, but the costs and complexities of sequestering CO_2 are not figured into most sales pitches for synfuel plants.

If the goal were really to use coal to help lessen our dependence on foreign oil, a better way to do it might be to push for the deployment of plug-in hybrid cars — so-called e-hybrids. An e-hybrid is essentially a regular gasoline-electric hybrid outfitted with larger batteries and an electrical outlet. You charge the car at night, when electricity prices are low, then drive it on electric power during the day. If the electric power runs out, you can switch to gas. Essentially, e-hybrids replace oil with electricity. Not only do they save on fuel, but they also have lower CO_2 emissions. E-hybrids are not pie-in-the-sky technology; ambitious backyard mechanics are already installing larger batteries and outlets in conventional hybrids and turning them into e-hybrids. According to Joseph Romm, former director of the Office of Energy Efficiency and Renewable Energy at the U.S. Department of Energy, by using typical grid power, the CO_2 emissions from an e-hybrid are 35 percent lower than from a regular hybrid and 65 percent lower than from a gasoline-powered car. An added bonus is that e-hybrids open the door to using cars as mobile generators: you might simply plug the car into the grid and feed power back to it. Instead of lighting our homes with coal plants, we could light them with power from the car in the garage. And that is exactly why many people in the coal industry would prefer to build billion-dollar coal-to-liquids plants. By keeping solutions to our energy problems as expensive as possible, Big Coal keeps out those pesky entrepreneurs who threaten to undermine the coal industry's interests.

Ultimately, the appetite for coal and the dangers of global warming are on a collision course. Since the world is obviously not going to kick the coal habit anytime soon, the only way to keep from overheating the planet is to find some way to reduce CO_2 emissions from coal plants — or to find a way to remove CO_2 from the atmosphere in equal proportion. There are lots of ideas about how to do this, from planting trees to deploying genetically engineered bacteria to help suck up the CO_2. But right now, there's only one method of getting rid of CO_2 from coal plants in quantities large enough to matter: pump it underground.

"The whole notion of a viable future for coal in a climate-constrained world hinges on the viability of CO_2 storage on a 'giga-scale,'" writes Princeton University professor and energy expert Robert Williams. Think of geologic carbon storage as an inverted smokestack: instead of pumping CO_2 into the sky, you pump it a couple of thousand feet underground, cap it off, and forget about it. There is nothing particularly radical about the idea. Oil companies have been injecting CO_2 into Texas oil fields to improve recovery for more than twenty years. In the North Sea, 6 million tons of CO_2 has been sequestered in gas fields under the sea floor. Globally, hundreds of millions of dollars are currently being invested in carbon storage research. It's an appealing idea: dig up coal, steal the energy, and then bury the nasty stuff down in the earth, where it won't bother anybody. It's a man-made carbon cycle, an industrial ecosystem.

The only place in North America where you can see what this

looks like is the Weyburn oil field in Saskatchewan, Canada. At first glance, the Weyburn field looks like any other oil patch: a dozen blue and yellow pumpers toiling on the prairie, heads going up and down like giant mechanical insects; plumes of dust visible in the distance as roustabouts in their beat-up 4x4s make their way from rig to rig, checking pipeline valves and pressure; a few scattered houses, mostly mobile homes, structures that attest to the temporary nature of every oil field settlement.

What makes Weyburn different is what goes on underground. Each day, about 240 million cubic feet of CO_2 from the Dakota Gasification Company plant in Beulah, North Dakota — a byproduct of turning coal into syngas — is piped 205 miles over the northern plains and under Lake Sakakawea until it reaches this old oil field on the Canadian side of the Williston basin. Here the CO_2 is pumped several thousand feet underground, where it flows through the porous rock, emulsifying and partly dissolving the crude, causing it to flow out of the reservoir more easily. Since CO_2 injection began four years ago, production in the oil field has increased by 50 percent, helping to offset the thirty-dollars-per-ton cost of capturing and storing the CO_2. When the reservoir is completely drained, it will be capped off, and the CO_2 will be locked in place. Over the first five years of the project (which was funded by a $28 million grant from the U.S. Department of Energy), about 6 million tons of CO_2 will be stored underground.

According to Jim Dooley, a senior staff scientist for Battelle, a global engineering firm involved in several carbon capture and storage projects, there are more than enough potential storage sites in North America to capture all the CO_2 emitted from U.S. industrial sources. Although some of those storage sites are in old oil fields and coal beds, the vast majority (about 95 percent) are in deep saline aquifers several thousand feet beneath the surface.

But carbon storage is a little more complicated than just drilling a hole in the ground and pumping in the CO_2. For one thing, the CO_2 first has to be captured from power plants, then compressed into a supercritical fluid to be injected underground. Although it is theoretically possible to capture carbon from an old coal burner, the cost of retrofitting the plant would be so high that most power companies would simply shut it down. With new coal plants, capturing and storing CO_2 from an IGCC plant would be much easier and cheaper than from a conventional combustion coal plant, in large part because the CO_2 could be removed from the syngas prior to combustion, rather than after, as it is at a conventional plant. (From the point of view of carbon capture, a syngas plant like the one in North Dakota, an IGCC power plant, and a coal-to-liquids plant are all similar.) Still, even at an IGCC plant, capturing and storing carbon is expected to raise the wholesale price of electricity by 20 to 75 percent. Using CO_2 to enhance oil recovery, as they do in Williston basin, would help reduce costs, but that is possible only in a few regions of the country.

The potential for carbon storage also varies enormously by region. Places such as Texas, Wyoming, and Illinois are well suited to it; the Carolinas and New England are not. Internationally, the United States and Canada have plenty of capacity; China and Western Europe have less; Japan and Korea have very little.

In any case, widespread deployment of carbon capture and storage would be a huge engineering project. To make any significant dent in the amount of CO_2 released into the atmosphere, thousands of underground CO_2 reservoirs would have to be drilled and maintained in coal-burning nations. For example, in the Weyburn field, which is currently the largest sequestration project in the world, during the project's twenty-five-year life span, 25 million tons of CO_2 will be stored underground. That sounds like a lot, but it's only about as much CO_2 as Georgia Power's Plant Scherer releases in a single year. To have a reasonable chance of stabilizing the climate, scientists believe we'll have to bury about one gigaton — a billion tons — of carbon by 2050. The scale of such an undertaking is mind-boggling. Lynn Orr, director of the Global Climate and Energy Project at Stanford University, estimates that the pipelines, pumps, and other infrastructure required to push this much carbon underground would equal the current infrastructure of the entire global oil and gas industry.

How safe is it? Some environmentalists have compared burying CO_2 to burying nuclear waste, which is unjustified. CO_2 is not radioactive, nor is it toxic by any conventional definition. But burying

CO_2 also is not risk-free. The release of toxic heavy metals is one concern. When CO_2 is pumped into underground aquifers, it makes the water more acidic. Acidic water dissolves certain types of rocks, releasing minerals and nasty stuff such as arsenic and uranium, which can then seep into drinking water supplies.

Leakage is a problem as well. CO_2 is buoyant underground and can migrate through cracks and faults in the earth, pooling in unexpected places. This is troublesome because CO_2 is an asphyxiant — in concentrations above 20 percent it can cause a person to lose consciousness in a breath or two. At Mammoth Mountain, California, where CO_2 naturally seeps up through the ground, skiers have died when they tumbled into pockets of CO_2 that had built up in hollows in the snow. In theory, you could enter a basement flooded with CO_2 and, because it's an invisible, odorless gas, you would never know it's there. If present in high enough concentrations, the gas could kill you. In 1986, near Lake Nyos in Cameroon, CO_2 from an underlying magma chamber seeped into the water, creating an enormous deposit of CO_2 at the bottom of the lake. One night, something — a landslide or heavy winds — disturbed the CO_2-laden water, causing it to rise to the surface, where 300,000 tons of CO_2 were released in a misty cloud. The cloud flowed down through two valleys, asphyxiating 1,700 villagers and thousands of cattle.

According to a recent report by the U.S. Department of Energy's National Energy Technology Laboratory, "the risk of large, catastrophic releases of CO_2, such as Lake Nyos and Mammoth Mountain, are virtually nonexistent for geologic sequestration." Still, it's far from clear how Americans will feel about living above giant pools of CO_2. Will people view it as a smart underground recycling program or as a dangerous way of burying their problems? For Big Coal, geologic carbon storage is nothing more than a high-tech garbage dump, a way of getting rid of a bothersome waste product. They will obviously be interested in doing it as cheaply as possible. Who will make sure it's buried responsibly and safely? Given Big Coal's history of regulatory abuse, will people trust federal and state officials to make sure that power companies don't cut corners? Of course, compared to the risks associated with global warming, the dangers of living above a CO_2 reservoir may seem small indeed.

Chapter 10

The Frontier

IN WESTERN CHINA, global warming is not a theory or an article of faith; it's a slow-motion tragedy. Xinjiang, a far western province that borders India and central Asia, is the poorest region in China, a remote frontier of open basins, snowcapped mountains, and big desert skies. It's the region of the ancient Silk Road, the first global superhighway, where camels still roam and villagers believe the deserts are haunted by travelers who perished in the sand.

Civilization has always had a weak hold on the land here, but thanks to the warming climate, it's become increasingly tenuous in recent years. The deserts are expanding, creeping into towns and covering fertile valleys. Topsoil is blowing away in epic dust storms that choke Beijing and carry Chinese soil all the way to California. In the mountains, glaciers are melting, causing flooding and jeopardizing the region's long-term water supply. You can blame many of the problems on Mao Zedong, who encouraged the clear-cutting of forests, the overgrazing of farmland, and other environmental crimes. But this previous devastation has only left the region more vulnerable to climate change. In Qitai, a town in the agricultural hub of the region with a population of nearly 250,000, I met a local official who declared point-blank that if something was not done to stop the desertification in Xinjiang, the town of Qitai would cease to exist in the near future.

In Xinjiang, as in much of the rest of the world, vulnerability to the effects of global warming is a stepchild of poverty. One of the biggest challenges in China today is figuring out ways to bring the rural poor along in the economic boom that has done so much to improve the lives of people in coastal cities such as Beijing and Shanghai. Few regions of the country lag as far behind as Xinjiang, where the average annual income of a farmer is about $360 a year. Like Tibet, Xinjiang is one of China's autonomous regions, where an ethnic group is given limited control over the provincial government. Xinjiang is dominated by Uighurs, a Turkic people with a long history in the deserts of central Asia. The Uighurs are to Xinjiang roughly what the Native Americans were to the American West — a colorful, nomadic civilization that is on the verge of being steamrollered by industrial power.

Over the past several decades, there have been numerous bloody clashes between the Uighurs, who are Muslim, and the Han Chinese, who dominate the central government. In part to quell these uprisings, Beijing has poured billions of yuan into economic development projects in the province, hoping to create jobs and raise the standard of living. Xinjiang now has the biggest ketchup plant in the world as a result of this initiative. The capital city, Ürümqi, which was little more than a desert crossroads twenty years ago, has a population of 2 million today. In recent years, Xinjiang has become increasingly important to China's future because of the region's large untapped oil, gas, and coal reserves. Just a few months before my visit, China's first east-west pipeline opened, carrying natural gas from Xinjiang to Shanghai, where it is burned instead of coal for heat and electricity.

In dirt-poor Xinjiang, progress is still fueled by black rocks; you see coal plants on the outskirts of most towns and cities. It doesn't take much effort to imagine what this region will look like in the near future. Sometimes, driving across the World Bank–funded highway that connects Ürümqi to the more remote areas, you can look across rolling sand dunes created by the desertification of agricultural land and see chugging smokestacks in the distance. It's easy to get depressed by such sights, to imagine that in twenty or thirty years, the fragile beauty of Xinjiang will be crushed by rampant industrialism.

But hope appears in unlikely places. One afternoon, I visited an Uzbek (another ethnic minority in Xinjiang) village where a farmer had found a particularly interesting solution to his energy problems. I was traveling with Dan Dudek, chief economist for Environmental Defense, and several colleagues, both Chinese and American. Dudek was in Xinjiang to set up a carbon trading program (more about that later) and was interested in visiting with people — farmers, villagers, industrialists — who had found creative ways to reduce carbon emissions.

We traveled down a long dirt road through field after field of wheat and cotton, all of it divided into small plots and worked by farmers with donkeys. At the end of the road was the village, maybe thirty houses in all, each little more than a mud hut with an enclosed courtyard, where the farmers kept their donkeys and chickens. Our Chinese guides led us to one of the most substantial houses in the village, with a cheerful red door and a clay tile roof. The farmer who lived there, a dark, leather-skinned Uzbek in his forties, was waiting for us inside. He showed us into his kitchen, a small room with an old stove, a metal-frame bed, a few rotting heads of cabbage, and not much else. Most cooking in villages is done with coal or (rarely) wood, but here the farmer turned the knob on the stove, and a clean blue flame jumped up: gas! Where did he get the gas? He pointed to the small pipe that went across the wall, through the next room, and into the small attached barn. We walked out to the barn, and he showed us a hole in the floor that was covered with a heavy rock: a methane digester. A methane digester is little more than a sealed concrete box filled with animal and human waste. As the waste breaks down under normal organic processes, the methane is captured and piped away. In effect, this farmer had replaced coal with crap. By the proud look on his face, you would have thought he'd invented the Internet.

During the next half-hour or so, we got the full tour of his eco-efficient house. He showed us the jury-rigged solar water heater he had built and installed on his roof, a low-wattage fluorescent bulb that hung from a wire in the middle of his home, and a Coleman lantern–style gas lamp (the latter two his only sources of light at night). What made this so impressive was not just that he had done so much with so little but also how proud of it he was. He was not just a farmer; he was a budding entrepreneur.

As we walked out to our vehicles a half-hour later, Dudek observed, "All we need is about one billion more people like him, and the world's energy problems will be solved."

Depending on what statistics you cite, China may or may not be the fastest growing economy in the history of the world. But it is without question the world's premier coal junkie. The Chinese burn less coal per capita than the Americans, but in terms of sheer tonnage, they burn twice as much. Seventy percent of the nation's energy comes from coal, and Chinese leaders have made it clear that economic growth is their number one priority. They see themselves as following essentially the same development path that the West took: get rich first, clean up later.

For China, this strategy represents an enormous gamble. Workers die every day in unsafe Chinese coal mines. Outdoor air pollution kills hundreds of thousands of people each year. Acid rain is falling on a third of the country, crippling agricultural production. Not surprisingly, pollution is becoming a source of political turmoil. A few weeks after my visit to Xinjiang, fifteen thousand people in a village south of Shanghai took to the streets, throwing stones and overturning police cars, demanding that a pharmaceutical plant that was polluting the local river be shut down. A few months earlier, in the nearby city of Dongyang, six police officers were killed in riots over pollution from a pesticide plant. To Chinese leaders, it has become increasingly clear that if economic growth is seen as a mechanism that allows the rich to exploit and poison the poor, there could be trouble. "We are convinced that a prospering economy automatically goes hand in hand with political stability," Pan Yue, China's outspoken deputy minister of the environment, said in an interview in 2005. "And I think that's a major blunder. The faster the economy grows, the more quickly we will run the risk of a political crisis if the political reforms cannot keep pace." Pan believes that pollution and ecological collapse, particularly in western China, will lead to 150 million environmental refugees flooding into eastern cities. Where will they go? What will they do for work? Whom will they blame for their troubles? "Chinese leaders do not fear democracy," Husayn Anwar, an executive at BP who has spent many years in China, told me. "They fear chaos."

This is not just an internal political problem. A political crisis in

China, whether it's brought on by too many dead Chinese coal miners or too many toxins in the Songhua River, would have devastating consequences for the rest of the world as well. For the West, the economic and geopolitical risks of a destabilized China are compounded by the longer-term dangers of global warming. The coal-fired power plants that China will build between now and 2012 will generate more than twice the amount of greenhouse gases that the Kyoto Protocol signers agreed to cut by 2012. By 2025, if China continues down the path it's on, it will overtake the United States as the largest greenhouse gas polluter in the world. In other words, if China doesn't figure out a way to avoid the coal-burning development path of the West, the chances that the world will be able to meet a warming target of 3.5 degrees Fahrenheit — the point beyond which the risk of abrupt climate change begins to increase dramatically — are virtually zero.

This is not to suggest that China has done nothing to clean up its act. On the contrary, it has done a lot. It has banned the use of coal for heating and cooking in cities such as Beijing and Shanghai; it has moved coal-fired power plants out of urban areas and replaced them with natural gas; it has tightened energy efficiency requirements on new buildings, factories, and consumer products; and it is in the process of building the largest offshore wind farm in the world. In some ways, China has already leapfrogged ahead of the United States. In 2005, the central government announced that 15 percent of electricity in the country will be generated by renewable sources by 2020. (The same year, a similar provision was stripped out of the U.S. energy bill.) China also has passed vehicle fuel-efficiency standards that equal or surpass anything in the West. But even with all these measures, China still has a long way to go.

Given what's at stake, it's obviously in the best interests of the West to help China deal with its environmental problems, especially when it comes to addressing global warming. We all live on the same planet, after all. If the world is going to have any chance of reducing CO_2 emissions to a level that stabilizes the climate, China will have to be a big part of the conversation. At the same time, nobody is going to strong-arm Chinese leaders into mounting a campaign to cut CO_2 emissions if it means risking economic growth and political stability. So the question becomes, How does the West help

China evolve quickly from massive industrial polluter to a leaner, greener economic machine?

From the Bush administration's point of view, the answer is simple: better technology. In July 2005, the administration announced the Asia-Pacific Partnership on Clean Development and Climate. It is essentially a trade pact among six nations — the United States, China, India, Australia, South Korea, and Japan — that is supposed to help speed the deployment of new, more sophisticated power plant technology in China and India, with the aim of making these nations' economies more efficient and reducing air pollution, including greenhouse gases.

When the pact was announced, it was seen by some as a not-so-subtle attempt to undermine the Kyoto treaty, offering big coal-burning nations a way to look like they are acting progressively to deal with global warming while in fact doing nothing much at all. The Bush administration denied this. At the press conference announcing the agreement, James Connaughton, head of the White House Council on Environmental Quality, emphasized that the partnership would be "complementary" to the Kyoto Protocol and was not intended to compete with it. Few believed this claim, including Republican senator John McCain. "The [Asia-Pacific] pact amounts to nothing more than a nice little public relations ploy," he told the writer Amanda Griscom Little. "It has almost no meaning. They aren't even committing money to the effort, much less enacting rules to reduce greenhouse gas emissions."

In early 2006, a few months after McCain spoke, the Australian government pledged to contribute $75 million over the next five years to jump-start the trade pact's technology fund, while President Bush promised to request $52 million for the fund from Congress. It's a start, but it's scarcely enough to buy new hardhats for Chinese coal miners, much less transform the global energy industry. And the members of the trade pact remained adamantly opposed to any rules that would limit greenhouse gas emissions.

Of course, Big Coal doesn't need a trade pact to encourage the industry to sell its wares and expertise overseas. In 2003, I happened to be on a flight from Chicago to Beijing that was full of beefy, well-groomed, middle-aged men in polo shirts and jeans. Many of them seemed to know one another. A few hours into the flight, while I was waiting to use the bathroom, I struck up a conversation with the guy behind me. As it turned out, he worked for a company that sold coal-washing equipment and was traveling to China to negotiate a deal with a Chinese coal company. He introduced me to several of his friends. They too worked in coal-related industries. In fact, the plane might as well have been chartered by Big Coal. One of the guys dubbed it "the Rust Belt Express." In 2005, Peabody Energy opened an outpost in China, and other American coal companies are surely not far behind.

Despite the dismissive reaction to the trade pact by some, it must be said that any deal that helps China jump ahead to better, cleaner technology is a step in the right direction. If the Chinese install first-rate scrubbers on their new coal plants, it will dramatically improve air quality in many regions, just as it has in some regions of the United States. In addition, companies such as GE and Bechtel are eager to begin building IGCC plants in China, which would be a huge leap forward.

But the administration's faith in technology as the magic-wand solution to every problem has a number of significant flaws. For one, much of the so-called new technology that the coal industry is trying to pawn off on China is actually second-rate stuff that is already out of date. (The coal industry, of course, isn't the only business that sees China as a vast dumping ground for old goods.) In addition, the uptake of truly important technology, such as IGCC, is slowed by legal and intellectual property rights issues that trade pacts such as the Asia-Pacific Partnership have only just begun to address. But the real problem is that "new technology" is usually defined as anything that helps encourage China's dependence on coal. Nobody who is involved in the partnership — as of this writing, anyway — is touting new solar panels or talking about how more efficient building design can help eliminate the need for new power plants. In short, the talk about better technology often seems to be more about promoting the interests of the coal industry than it is about genuinely addressing the problem of global warming. It leads to obvious suspicions that the trade pact is really just another part of Big Coal's long-running campaign to encourage consumption and clean up coal's image as a viable fuel for the future. "It's the

FutureGen road show," one U.S. environmentalist dubbed the partnership.

Indeed, encouraging a global dependency on coal is a key part of the industry's agenda — and not just because it gives Big Coal new markets in which to sell its goods and services. This is particularly true of China, the nation that undoubtedly poses the most direct threat to America's future economic dominance. Among other things, China's coal habit gives Big Coal supporters in the United States moral cover to argue that taking any meaningful action to limit CO_2 emissions will, as President Bush put it in 2005, "wreck our economy." After all, why should America put itself at an economic disadvantage to save the planet if the Chinese won't stop burning coal, too? By raising the specter of the Chinese threat to America's status as the world's only superpower — a threat that is unreasonably potent in the American subconscious — Big Coal supporters are able to equate coal consumption with American prosperity, freedom, and military power.

In fact there is more than a whiff of hypocrisy in America's efforts to help China clean up its act, especially when it comes to global warming. "I hear about it every time I visit," says Douglas Ogden, head of the China Sustainable Energy Program of the David and Lucile Packard Foundation. "They even joke about it with me, it's so obvious and absurd." Chinese leaders understand very well that the reason global warming threatens the stability of the planet today is because the industrialization of the West — those 150 blissful years of burning fossil fuels — loaded up the atmosphere with CO_2. They also are quite aware of the fact that the average American is thirty times richer than the average Chinese, and they don't hesitate to remind people of it. As one Chinese delegate involved in negotiations over the Kyoto Protocol put it, "What [developed nations] are doing is luxury emissions. What we are doing is survival emissions."

Ürümqi is often cited as one of the ten most polluted cities in China, but when I visited in mid-June 2005, the haze was no worse than it is in Pittsburgh during the middle of the summer. (Air pollution in China is usually worst in the winter, when coal is burned in home

ANNUAL PER CAPITA CO$_2$ EMISSIONS

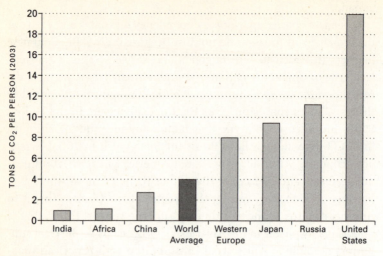

Source: EIA

furnaces for heat.) But it was hot. By 11:00 A.M., the temperature was already above 90 degrees Fahrenheit. Women strolled the streets with umbrellas, children huddled under poplar trees on the main boulevards, and decrepit air conditioners hummed in the windows of every apartment building.

Nobody I talked to had any illusions about what was going on. On the way to a meeting with Chinese officials on the first day of my visit, I rode in a van beside Wang Yijian, a vice director of Xinjiang's Environmental Protection Bureau (EPB), the equivalent of a state environmental agency in the United States. I asked him if the heat was typical of early-summer weather in Xinjiang. He shook his head and said, "Summer used to arrive in late July. Now it arrives in early June. Winters are getting warmer, too. The low temperature in winters used to be minus 30 degrees [C]. Now it is only minus 20 degrees. We see these changes very clearly."

"Why do you think this is happening? Global warming?" I asked, expecting the same kind of qualified response that I might get from a government official in the United States.

"Yes, yes," Wang said, nodding his head for emphasis. "People here are very concerned about it. Seven or eight years ago, no one needed an air conditioner in Ürümqi. Now it is necessary. It is of course expensive, and it means burning more coal to create the electricity to power those air conditioners. It is a difficult situation. More power, more pollution, more power . . ." He made a kind of circling motion with his hand, suggesting an ever-upward climb.

I nodded. I knew what he meant: *the greenhouse spiral.*

A few minutes later, we arrived at the headquarters of the Xinjiang EPB, a typically unadorned concrete box in downtown Ürümqi. We all sat in a conference room around a big rectangular table — Dudek and his colleagues on one side; Niu Xiaoping, the head of the Xinjiang EPB, and her colleagues on the other. We had met Madame Niu the night before when she'd greeted us at the airport. She is a short, middle-aged woman with prominent teeth and a fiercely intelligent manner. She and Dudek had bonded during the bus ride from the airport to the hotel: he'd told her about his two adopted Chinese daughters; she'd told him about the genetically altered colored cotton farmers are growing in Xinjiang. This morning, she was clearly eager to hear Dudek's presentation.

Dudek has been working in China for more than ten years. During that time, he has eaten pigs' ears and roosters' testicles; he has thrown back countless shots of *baijiou,* a sorghum-based liquor favored by the Chinese, at hundreds of ceremonial lunches; he has been honored in the Great Hall of the People for his environmental work in China. He knows how important it is to observe the rituals of politeness with the Chinese, how to be at once blunt and solicitous, independent and respectful. At fifty-eight, with silver hair and mustache, dressed in a blue suit and red tie, Dudek looks more like a Wall Street banker than an environmentalist. But at heart, he is a radical capitalist, a ThinkPad-toting free market zealot. So what's he doing in China? "There aren't many leverage points in global warming," Dudek told me a few days earlier on the flight from Beijing to Ürümqi. "China is one of them. Because the country is so big and growing so fast, there is real possibility for change here. This is where the future is being built."

In the United States, Dudek is best known as one of the primary architects of the 1990 amendments to the Clean Air Act (also known as the Acid Rain Program). In 2002, *The Economist* called the sulfur dioxide trading program "the greatest green success story of the past decade." It was the success of the 1990 amendments, in fact, that led to the idea of emissions trading being incorporated into the Kyoto Protocol as a way to reduce global emissions of CO_2. Like the 1990 amendments, the trading program agreed to in Kyoto in 1997 (and expanded in subsequent revisions to the treaty in Bonn and Marrakesh in 2001) gives CO_2 caps, or quotas, to each developed nation that is party to the treaty, then allows them to meet those caps either by reducing CO_2 levels or by purchasing allowances from other countries who have reduced their CO_2 emissions below their caps.

But the trading program agreed to under Kyoto also differs from the U.S. sulfur dioxide program in a number of ways, including the use of the so-called Clean Development Mechanism (CDM). The CDM allows developed countries to get credits for investing in projects in developing countries that offset greenhouse gas emissions, where the cost of making reductions is often much lower. For example, a Danish power company can earn carbon allowances by setting up a system to capture methane from a rotting landfill in Brazil or by erecting wind turbines in China that offset dirty coal power. Although CDM projects in many nations have been slowed by legal complexities and bureaucratic paper shuffling, the program has enormous potential. According to the World Bank, as much as $12.5 billion will be spent on CDM projects in developing nations by 2012.

Right now, carbon trading and CDM investment is still a very new idea, but it's catching on quickly. Although the carbon market won't take off until Kyoto goes into force in 2008, the European Union has already set up a thriving carbon exchange. In 2005, the first full year of trading in the EU, the carbon market grew to $7 billion. By the time the Kyoto Protocol kicks in, the market is projected to be close to $50 billion. Carbon trading has already created a new industry of financial advisers and brokers; now major investment banks, such as Goldman Sachs and Merrill Lynch, are getting into the game. Many investors believe that carbon will soon be the largest commodity market in the world.

"To me, the beauty of carbon trading is that it has the potential to harness a primal human impulse — greed — and redirect it toward saving the planet, rather than destroying it," Dudek explained to me. "It's also about empowering people, getting them involved. One of the problems with global warming is that it can seem so huge, so overwhelming, so much about saying no, about shutting things down, about denial and hardship. It's easy for people to get depressed about it, to think it's beyond their control, to say, 'Forget it. I'm going to go watch *Desperate Housewives*.' To me, the key questions are not 'How do we pass laws to mandate the construction of IGCC plants?' or 'Should we build more nukes?' but 'How do we get people engaged? How do we get people to think differently about this problem? How do we inspire creativity?'"

China is a signatory to the Kyoto Protocol as a developing nation, which means it is not bound by the treaty to make reductions in greenhouse gas emissions. But the country can still benefit from CDM investments from developed nations. In Xinjiang, Dudek's goal was simply to set up a demonstration project to show the Chinese how it all works. He picked Xinjiang for a number of reasons. Economically, the region is enough of a blank slate that it might still be possible to change the vector of industrial developments ever so slightly, demonstrating that there are other paths to prosperity than simply to burn coal, erect high-rises, and build ever-larger ketchup factories. Dudek's goal in the coming months would be to assemble a portfolio of microscale carbon projects — no-till farming, crop and forest sequestration, perhaps some methane gas capture at landfills — that are cheap and easy to do, will contribute to better land management, and, most important, will help put money in the pockets of poor farmers. He would then market the portfolio to potential buyers — perhaps an American energy company looking to offset carbon emissions and buy a little goodwill with the Chinese at the same time, or, even better, to a Chinese energy company eager to demonstrate its progressive values and commitment to environmental principles. But for any of this to work, Dudek needed to get provincial officials in Xinjiang to commit to working with him in partnership on the project, not necessarily an easy sell given the financial and cultural complexities of the deal.

A few minutes after we arrived at the EPB, Dudek opened his ThinkPad and fired up his PowerPoint presentation. Although he does not speak Chinese (he uses a translator), his manner was relaxed and thankfully short on catastrophe, guilt, and fear-mongering. Instead, he focused on market fundamentals, explaining how carbon trading is an outgrowth of the Kyoto Protocol, and how China, in the long run, could benefit from thinking of greenhouse gases as a valuable commodity, rather than merely as invisible waste that is pumped into the atmosphere. It was all boilerplate stuff and wouldn't have been worth noting if this meeting had been taking place in, say, London. But here in this small, dark room in one of the poorest provinces in China — a nation that is, after all, still nominally run by the Communist party, where the legal system is at best shaky, and where transparency and accountability are new ideas to many Chinese bankers and businesspeople — it all sounded downright revolutionary. Watching the faces of the Chinese officials in the room, I could see their world being turned upside down. *Global trading? CDM mechanisms? Cash for carbon?* Goodbye, Marco Polo; hello, Warren Buffett!

"We feel this is a very promising approach to our problems in Xinjiang," Madame Niu said shortly after Dudek wrapped up his presentation, her eyes bright with enthusiasm. "It may be an American idea, but we believe it has Chinese characteristics."

In 1991, on his first trip to China, Dudek participated in a workshop on air pollution at a resort in Shandong, a coastal province near Beijing. Each morning, he went for a walk in the nearby countryside. "The thing that struck me was not the amount of pollution in the air," Dudek recalled. "It was the silence. There were no birds."

Empty skies in China are a poignant reminder of Mao's brutal legacy. "Man must conquer nature," he proclaimed. Mao identified sparrows as one of the "four pests" (rats, flies, and mosquitoes were the others) because they ate grain in the fields, in effect stealing from workers. He commanded people to kill the birds, which they did, often by banging gongs whenever the birds tried to land, until they dropped dead from exhaustion. It was all part of Mao's relentless drive to transform China overnight from an agrarian society into an industrial powerhouse. In 1958, millions of Chinese were mobilized to clear forests, build dams, and smelt steel in backyard furnaces. In *Mao's War Against Nature*, author Judith Shapiro estimates that 10 percent of the forest cover in China was cut down in a matter of months. Erosion ruined valleys and flooded rivers. Crops failed on a massive scale. Between 1958 and 1961, 30 million Chinese starved to death.

Today an entire generation of Chinese has grown up disconnected from the natural world. "I didn't know what nature was until I was thirty years old," Wang Yongchen, a well-known Chinese activist and journalist, told me in her cramped, high-rise apartment in Beijing. "Then in 1989, I took a trip to western China, and I discovered animals, trees, birds. They were so happy together! I did not know such things existed in my country. After that, I decided to devote my life to trying to help Chinese people reconnect with nature again." Her popular radio show, *Read Green*, broadcast on China National Radio, is a weekly ecological adventure in which she takes listeners to places such as a bee farm or in search of a rare species of lotus. Simple as it sounds, such a show would have been impossible to imagine in China even ten years ago.

In some ways, the birth of China's environmental consciousness parallels Dudek's own. He grew up in a four-room tenement in West Springfield, Massachusetts. His father worked for a subsidiary of Standard Oil in a factory that made oil burners and gasoline pumps. They were working-class poor: no phone, no TV, no backyard. In the late 1960s, after attending Cornell University for a few years, Dudek enlisted in the army and was stationed at a base in Korea, where he sometimes had the unlikely job of riding around on garbage trucks to make sure people didn't smuggle items off the base in them. He became fascinated by the way a GI's trash — a discarded Coke can, for example — became a valuable object to a poor villager. The insight changed his life. When Dudek returned to the United States, he began studying resource economics — how commodities such as air and water are apportioned — and eventually earned a Ph.D. from the University of California, Davis.

In California, he worked for the U.S. Department of Agriculture on water quality issues, then spent a few years teaching at the University of Massachusetts Amherst before being recruited to join the Environmental Defense Fund (now called just Environmental Defense), an organization founded to propose market-based solutions to environmental problems. Not long after he arrived, he found himself in the White House, working on air pollution legislation.

Cap-and-trade was not Dudek's idea. It had been floating around in academic circles since the 1960s and had been used in an experimental air pollution control project in Southern California, as well as in the Montreal Protocol, the international treaty that limited ozone-harmful chemicals in 1987. But Dudek and Joe Goffman, his colleague at Environmental Defense, were instrumental in shaping the new legislation being pushed by President George H. W. Bush to solve the problem of acid rain. The key innovation was allowing companies that cut sulfur dioxide beyond the required amount to sell their "extra" tonnage, in effect creating a giant open market for emissions reductions. It worked even better than Dudek expected. By 1995, sulfur dioxide emissions were 22 percent below what was required by law, while compliance costs were 75 percent less than the EPA had originally predicted.

The success of the program helped inspire an ideological revolution in how to deal with pollution. According to previous environmental dogma, markets were part of the problem, not part of the solution. (David Brower, the longtime director of the Sierra Club, called economics "a form of brain damage.") The Acid Rain Program proved otherwise. Eileen Claussen, who worked on the implementation of the legislation while she was at the EPA and is now director of the Pew Center on Global Climate Change, calls the sulfur dioxide trading program "a paradigm shift in how we deal with air pollution in this country and around the world." Before long, markets were erected to help protect water supplies, conserve habitats, and manage fisheries. The introduction of marketable quotas is widely credited with helping to restore the Alaskan halibut fishery, which was on the verge of collapse in the late 1990s. In New Zealand, a tradable quota system now covers ninety-three fish species.

After the success of the 1990 amendments, Dudek could have cashed in his chits and landed a cushy inside-the-Beltway job. Instead, he went to China. "I consider myself a blue-collar environmentalist," he told me one night as we were riding through Tiananmen Square in the back of a taxi. "I like to be out in the world, actually trying to fix things. A lot of environmentalism in the U.S. is all about giving good paper, and I have zero interest in that." Dudek has spent the better part of a decade setting up a sulfur dioxide trading program in China modeled on the one in the United States. After experimenting with the idea in a few pollution-conscious cities in eastern China, officials in Beijing were impressed enough to include a directive in the eleventh Five Year Plan, the government's official roadmap for the future, to implement emissions trading more broadly in the coming years.

Still, in regard to carbon trading, China has been slow to jump on the bandwagon. The nation lags behind Brazil and India, which have already become adept at packaging and marketing carbon projects to Western nations. In part this reflects the more complex political system in China. Before Chinese officials throw open the gates to carbon trading or court CDM investment, they are trying to sort out their priorities: Who gets the benefits? Which regions should be targeted first? Which sectors of the economy will be allowed to participate? China also has lagged behind because there is still a lot of distrust in the Chinese legal and regulatory systems and their ability to verify emissions reductions and other data. "Carbon investors, like any other kind of investors, like easy, safe, slam-dunk deals," Dudek explained. "China is still figuring all this out. It is not a slam dunk."

Much criticism has been leveled at the role the World Bank plays in carbon trading, especially in developing nations. From a climate perspective, the World Bank's role has been anything but pure. Although it is by far the largest facilitator of carbon trading in countries such as Brazil and India, it is also the prime backer of fossil fuel development. Between 1992 and 2004, the bank approved $11 billion in financing for 128 fossil fuel extraction projects in forty-five countries. In short, the World Bank is playing both sides of the fence, and the net result is that instead of pushing CO_2 levels down, it is pumping them up.

But if there is a fundamental flaw in carbon trading, it is not in the power of the market to drive change or in the dealings of the

World Bank. Rather, it is in the character of the politicians and regulators who set up the playing field and make the rules. In the United States, the sulfur dioxide trading program succeeded because there were precisely defined rules, firm caps, and significant penalties for exceeding the caps. In contrast, President George W. Bush's Clear Skies Initiative failed to attract much support. Not only did it not limit CO_2 emissions but it was poorly designed: it had loose caps, a long timetable for achieving goals, and it included a toxin, mercury, that accumulates in local "hot spots" and is therefore not suitable for a national trading program.

A poorly designed trading system — whether it's for carbon or halibut fishing rights — is little more than an expedient way of appearing to do something about a problem while really doing very little. To succeed, regulators and politicians have to get the rules right, which, in the case of carbon trading, means striking a balance between what's good for the fossil fuel industry and what's good for the rest of us. For anyone who has had a close look at the political power of Big Coal in America, let alone China, this is a high hurdle indeed.

Still, even among hard-core activists, the idea of using the market to address complex environmental problems has won a lot of support in China. "I think the free hand of the market is much better than the heavy hand of dictators," said Dai Qing, a well-known dissident and author who served ten months in prison for her criticism of the Three Gorges Dam, the controversial project on the Yangtze River that has forced the resettlement of more than one million Chinese villagers. We were sitting in a hotel lobby in Beijing, and as she spoke, she often glanced over her shoulder. Despite the new openness, the secret police were still keeping close tabs on her. "My friends who are antiglobalists disagree with me," she said. "But I believe that if anything is going to save China from an environmental catastrophe, it's capitalism."

After visiting a wind farm and a landfill near Ürümqi, I flew with Dudek and his colleagues to Hetian, an ancient trading outpost on the Silk Road and now one of the poorest regions in Xinjiang. It was after dark when we arrived in the city of Hetian, and the broad main boulevard was full of people walking, riding bicycles, and driving donkey carts. There were very few cars. As we rode in a taxi from the airport to our hotel, Zach Willey, a carbon sequestration expert with Environmental Defense who was traveling with us, noticed that except for the streetlights, the city was dark — no lights in any of the apartment buildings, no neon, no restaurant lights. "Poverty," Dudek explained. "No one can afford to turn the lights on."

When we arrived at our hotel, we stashed our bags in our rooms and walked to an outdoor bar beside the hotel to get a beer. As it happened, the U.S. Senate had just passed the Energy Policy Act of 2005 that day. I'd received the news in several e-mails from friends in Washington. When I mentioned the bill to Dudek, he rolled his eyes. "Energy politics in America has become such a cynical game," he said. "It's all about the energy companies lining up at the trough and pigging out. That's one thing I find so inspiring about working in China. They've got big problems here, but at least they're trying to find solutions and experiment with new ideas. They are very pragmatic about it. If something works, great. If not, they move on to another idea."

Was it concern for the earth's climate or simply the prospect of making money that was driving China's interest in carbon trading? I wondered. "I think people here in Xinjiang have real concerns about global warming and about what it could mean to their way of life," Dudek told me. "But when you get down to it, it's mostly about money. But so what? That's the great thing about carbon trading. Purity of heart is not an issue. As long as the amount of carbon that's going into the atmosphere gets reduced, who cares what anyone's motives are?" He took a swig of his beer, then added, "I think people sometimes view my work in China as simply an experiment to see how you can apply market economics to a developing country. That is ridiculous. If China can't get a handle on its environmental problems, the consequences for the country — and for the planet — could be catastrophic."

The next morning, we drove for several hours along the edge of the great Taklimakan desert. (Taklimakan is variously translated as "if you go in, you won't come out" or "place of no return.") Sand dunes rose up along the road; the wind blew the sand in drifts across the road. We passed through oases, where tall poplars lined the road, villages and gardens thrived, and rivers mysteriously appeared then disappeared. The highway was rough and crowded with everything from coal trucks to Audis. Finally, we arrived in the cool, shady Uighur village of Ao Yi Tuo Ke La Ke, about one hundred miles northeast of the Afghanistan border, where fields of beautifully tended vegetables and walnut and pomegranate trees grew. It was as picturesque as Tuscany, except there were no hills and no villas. We drove to the edge of the village, where the oasis met the desert, then walked into a field of tall, slender anabasis shrubs (*Anabasis ammodendron*) that had been planted in the sand dunes to keep them from moving in and destroying the oasis. According to our Chinese hosts, the shrubs (native to the region) had saved the village by stabilizing the soil and breaking the wind. Could they help do the same thing for the climate?

If the idea of carbon trading is controversial, the use of biological sequestration is doubly so. Biological sequestration, or "carbon sinks," refers to measures that help increase the earth's natural ability to store carbon, usually either in plants or in soils. As plants grow, they absorb CO_2 from the atmosphere through photosynthesis, eventually storing carbon in wood, leaves, and soils. Globally, the destruction of forests and land-use changes are a major source of greenhouse gases, responsible for more than 20 percent of CO_2 emissions. In contrast, large-scale tree planting can help suck CO_2 out of the atmosphere. Similarly, small changes in agricultural practices, such as no-till farming and grassland restoration, can increase the amount of CO_2 retained in the soil. Although the amount of CO_2 that is sequestered may be small for any particular carbon sink, if you add enough of them together, they potentially can have a big impact on the balance of CO_2 in the atmosphere.

During the Kyoto negotiations, the U.S. negotiators fought hard to allow carbon sinks to be credited as legitimate sources of carbon offsets. Europeans fought equally hard against them, viewing carbon sinks as little more than an easy way for the United States, which has vastly more land for potential carbon sinks than Europe, to escape making real cuts in CO_2 emissions. In the end, the United States won the battle, although the subject remains controversial among both scientists and environmentalists. What is up for debate is not the fact that trees and soils absorb carbon, but how you measure and monitor that carbon, especially over the long term. In the view of some, the main problem with biological sequestration is not just that it's difficult to verify but also that it's a hopelessly small solution to a very big problem. According to one study, to sequester just 10 percent of the United States' annual CO_2 emissions, trees would need to be planted on one third of the nation's farmland.

Dudek isn't bothered by these numbers. He often talks about what he calls a "no regrets" strategy for dealing with global warming. That strategy entails finding solutions that can be implemented now to reduce greenhouse gas emissions, that don't require billions of dollars of investments, and, most of all, that have ancillary benefits that are worthwhile in and of themselves: poverty alleviation, ecosystem restoration, increased energy efficiency. These are all small steps and easy to write off as ineffectual given the enormity of global warming, but as the rise of China itself demonstrates, many people taking small steps can have a big impact. "I'm for anything that helps us think differently about global warming," he says. "A lot of people are fixated on cutting carbon dioxide emissions, which is fine, but it also narrows the conversation. Ultimately, what matters is what happens in the atmosphere, not what happens at any particular power plant."

Consider the rain forests in developing nations such as Brazil and Costa Rica. These enormous forests are huge natural air filters that absorb large amounts of CO_2, and yet they are being chopped down at an alarming rate by villagers and local businesspeople who are desperate to make a living any way they can. In a world looking for solutions to global warming, this makes no sense. At the United Nations conference on climate change in November 2005, a group of developing countries led by Papua New Guinea and Costa Rica came up with a novel solution: pay us, and we'll preserve our rain forests. Such a deal, if it were correctly structured, would have numerous benefits, not just for local residents but also for big coal-burning nations who were looking for cheap, innovative ways to reduce greenhouse gas emissions. "Right now, these rain-forest nations are providing enormous environmental services to the rest of the world — biodiversity, reduced greenhouse gas emissions — and they are not being compensated," says Nobel Prize–winning economist Joseph Stiglitz, who has championed the proposal. "From the viewpoint of economic efficiency, the best use of rain forests is to maintain them as rain forests."

The deal that Dudek was trying to put together on the anabasis plantation worked in a similar way, and is a good example of what he called a "bank shot" for carbon trading. In an ideal world, a company or nation looking for cheap carbon credits would pay the farmers to plant thousands of acres of anabasis in the desert, which would not only create a new revenue stream for the farmers but also help fight desertification and create habitat for desert wildlife. Even better, this particular variety of anabasis hosts a type of mushroom that the Chinese believe has a variety of medicinal properties, including the promotion of male "vigor." These mushrooms, which are very dear to the Chinese, could be harvested for profit in Dudek's carbon trading scheme, increasing the incentive to grow and protect these plants.

In terms of alleviating poverty, this was certainly a shrewd strategy. In Dudek's scheme, the revenue generated by selling carbon at $5 a ton would be roughly half a million dollars, an enormous sum in a region like Xinjiang. This would undoubtedly make a big difference in the lives of the farmers and others who participate in the deal. But it's hardly going to keep the Greenland ice sheets from melting. The carbon portfolio that Dudek was trying to assemble in Xinjiang might sequester 100,000 tons of carbon (equivalent to 367,000 tons of CO_2) over a period of seven years. That's roughly the amount of CO_2 that an average Chinese coal plant emits in a single month.

"The point here is not to solve global warming," Dudek said as we talked in the hallway of our hotel later that night. "That's dead-end thinking. That's the kind of thinking that leads to endless debate about what we should and shouldn't do. I've been hearing it for twenty years. How much longer can we talk about this? I've been in enough rooms where people whine about the evils of the coal or oil industry, who want to sit around and debate who is worse, Big Coal or Big Oil. You know what? Who cares? It gets you nowhere. What's important is to get people moving in the right direction. It's about leveraging creativity and energy, about working from the bottom up. It's about starting a revolution. We don't need Einstein here. We need Steve Jobs."

For centuries, Ürümqi was haunted by winds that blew down from a mountain in the center of town. Local residents came to know it as Ghost Mountain, because it seemed to be inhabited by the spirit of these deadly winds. The winds would sometimes blow for days at a time, covering the city with dust and sand, infiltrating into every mud home, and bringing life to a stop. Several hundred years ago, the residents of Ürümqi built a pagoda on top of the mountain and called it, according to one translation, Pressing Ghost Tower. The idea was that the pagoda would hold down the spirit of the mountain, and the dust storms would cease. The pagoda, which is still there, is a beautiful red and blue structure with a traditional curved roofline. It is one of the few architectural landmarks that escaped Mao's bulldozers. But beautiful as it is, it did not do much to reduce the dust storms.

About ten years ago, Chinese engineers came up with a better plan. First, they built an ultramodern sewage treatment plant on the top of the mountain (discreetly hidden on the back side, out of view of most residents). It was an expensive and ambitious project, largely funded by an influx of cash from Beijing. The sewage was pumped to the top of the mountain and treated in the plant. The wastewater, which was not clean enough to drink but was fine to use for watering plants, was piped into a vast gravity-fed irrigation system around the mountain. The residents of Ürümqi then began planting tens of thousands of pine, birch, and poplar trees on Ghost Mountain, watering them with the irrigation system. Today those trees are about eight feet tall, and Ghost Mountain has been transformed from a menace into a park. The spirit of the dust demon who once haunted the mountain has finally been suppressed.

On the last day of our visit to Xinjiang, our Chinese hosts drove us to the top of Ghost Mountain. We climbed up to a monument that had recently been built to honor the people who had planted the trees and built the park. The Chinese are extraordinarily proud of what has been accomplished here; it is as much a sign of their progress and sophistication as the Hoover Dam was for Americans seventy years ago. It is also a twist on Mao's quest to conquer nature: they had accomplished this not by exploiting nature, but by working with nature.

The view from the Ghost Mountain monument was remarkable. In the city below, we could see flocks of construction cranes — the national bird, some Chinese joke — swinging against the horizon. From here, the city looked chaotic, alive, metastatic. Like other desert cities — Las Vegas, Phoenix, Los Angeles — there is something both tragic and inspiring about seeing a city rise where none should be: is it a folly of human arrogance or a triumph of engineering? In the distance, I could see yellow-gray plumes rising from the candy-striped stacks of the city's coal plants. From my vantage point on the mountain, it was clear that global warming was not an "issue," but a manifestation of civilization itself, a byproduct of the very forces that bring life to the desert. I thought again about Henry Adams's visit to the World's Fair in 1893, and how the sight of the giant, snorting, coal-fired dynamos had both awed and frightened him. More than one hundred years had passed since then, and, as Adams predicted, the world had indeed changed irrevocably. But it still wasn't clear where our devotion to coal-fired power was taking us, or if the knowledge and sophistication we have gained by more than a century of coal burning will save us from destroying the very climate that gives us life.

The next morning, before we headed out to the airport, Dudek met with Madame Niu. "We have learned very much from you and your colleagues," she said. "We believe the project you propose could be of great benefit to the people of Xinjiang." Dudek responded in kind, and Madame Niu agreed to provide him with a memorandum of understanding within a few weeks. In other words, they were now in the carbon trading business together.

As we stepped onto the bus that would take us to the airport, I asked Dudek if he felt as though he had accomplished anything on this trip. "It's a microstep," he said in reply.

Epilogue

An Empire of Denial

THE LAST TIME America faced up to its energy problems, I was
a teenager living in Silicon Valley. I remember the long lines at the
gas stations and the sense of doom that hung over the world in
those days. My girlfriend and I used to sit on a blanket under the
stars and talk in an adolescent way about the troubles the planet
faced: overpopulation, nuclear war, air pollution, big IBM computers that would turn us all into drones, the rising military-industrial
complex. It was an odd time to be coming of age, a moment when
the future seemed bleak indeed.

Within a few years, everything changed. The gas lines disappeared, the yellow-brown smog that obscured the Santa Cruz Mountains began to clear up, and two guys named Steve built a computer
in their garage, named it after a fruit, and made a lot of people rich.
Before long, the Berlin Wall fell, the Soviet empire collapsed, and
the future I had imagined seemed a little less bleak.

Looking back on this time now, I realize what an extraordinary
period it was and how it has shaped my view of the way the world
works. Among other things, it transformed me into an optimist. It
taught me that change is necessary, that progress is possible, and
that huge and seemingly intractable problems can be solved in the
most unexpected and unlikely ways.

The three years I spent researching and writing this book tested
my faith. In the nineteenth century, during the heyday of coal, it was

easy to believe that the ladder of progress might extend forever, that civilization could be measured as a series of triumphs and discoveries in a never-ending spiral. Today we know that the earth's resources are finite, that reckless exploitation is not sustainable on a planet where a new city the size of Seattle is being built roughly every seven days, and that many of the advances of the industrial revolution came at great cost. In fact, the triumph over nature that the industrial revolution seemed to represent hasn't been a triumph at all. We just got lucky. We found ourselves living on a planet with lots of ancient decomposed plants and animals, we figured out how to mine them and burn them to create energy, and we built a great industrial empire. But now that era is coming to an end, and we have to figure out a new way to live. The question is, are we up to the task?

Right now, we remain in deep denial. Soldiers die in Iraq to protect our right to drive SUVs. We risk stirring up more Hurricane Katrinas just so we can crank up the AC. And when I say "we," I don't exclude myself. Writing this book has made me hyperaware of my own energy consumption, and I do what I can to lighten the load: I buy Energy Star appliances, burn compact fluorescent bulbs, ride my bike to the store instead of driving, purchase wind power from my electric company, buy as much food as possible from local farmers, and don't invest in coal company stocks or mutual funds that hold those stocks. But I am not living off the grid. I used a computer, not a quill pen, to write this book. "All left wing parties in the highly industrialized countries are at bottom a sham," George Orwell wrote in 1942, "because they make it their business to fight against something which they do not really wish to destroy."

That may be true. But we don't need to destroy our world, just reinvent it. In the coming decades, the great danger is not that the world will burn more coal — that's a given — but that we will burn it badly, cheaply, exploitatively. Instead of building modern IGCC plants that at least allow for the possibility of sequestering the CO_2 underground, we will throw up another generation of coal burners that will pump millions of tons of CO_2 into the atmosphere and accelerate global warming. Instead of strengthening laws against mountaintop removal mining, we will weaken them, allowing ever greater portions of Appalachia to be destroyed. Instead of shutting down dangerous mines, we will allow them to continue operating and more coal miners will die needlessly. Instead of helping developing countries leapfrog beyond coal, we will turn them into fossil fuel addicts like ourselves. But it doesn't have to be this way.

What can we do?

First, we must recognize that the world faces two enormous challenges in the coming years: the end of cheap oil and the arrival of global warming. Both are profound threats to our comfortable notions of civilized life. We should be grateful for the vast reserves of coal we have left and use them wisely, but the hard truth is that our bounty of coal is not going to save us from anything. At best, exploiting our coal reserves will buy us a decade or two of time and come at enormous expense, both in terms of the environment and public health and in terms of the billions of dollars that will be spent remaking a nineteenth-century fuel for the twenty-first century. In many ways, the world's coal reserves only make our energy problems worse, because they give us a false sense of security: if we run out of gas and oil, we can just switch over to coal; if we can figure out a way to "clean" coal, we can have a cheap, plentiful source of energy. In reality, however, facing the twin challenges of the end of oil and the coming of global warming is going to require reinventing the infrastructure of modern life. The most dangerous thing about our continued dependence on coal is not what it does to our lungs, our mountains, or even our climate, but what it does to our minds: it preserves the illusion that we don't have to change our thinking.

Second, it is important to see that the barriers to change are not technological but political. There is no reason we have to wait ten years (or longer) for FutureGen to come online before we build a near-zero-emissions coal plant. That could be done today with IGCC plants that capture and store CO_2 underground. Why aren't these plants being built? Because no law says that they must be and no customers are clamoring for them. Stockholders for the Southern Company or Peabody Energy aren't interested in underwriting efforts to maintain a stable climate. They're interested in maximizing the return on their investment. They will change only when we — or the politicians we have elected to speak for us — tell them they must, or when they see that it is in their economic interests to change (which is why shareholder actions that call attention to the financial risks associated with emitting large amounts of CO_2 show so much promise). That is the American way.

Third, we need to find ways to make the invisible visible. I mean this in the broadest possible sense. Big Coal has thrived largely because the costs of air pollution, miners' safety, devastated mountains, and global warming are invisible to us as consumers of electricity. There is no surtax on our electricity bills for charges related to increased asthma attacks in children or burial fees for victims of sudden heat waves related to global warming. If people could see behind their light switches, if they could trace the wires back to the power plant and the coal mine, if they could get an accurate accounting of what a kilowatt really costs us, would they still choose to burn coal? Some would, but some wouldn't. Either way, it would be much harder to maintain the fiction that electric power flows down to us from a giant golden bucket in the sky.

Making the invisible visible means changing this dynamic. This can be done in a variety of ways, including encouraging public service commissions to give electricity customers more choices of where they buy their power, requiring that electric power meters be placed in plain sight and be easy to read, and pushing for the development of a smarter electric grid that will allow people to monitor and control their power usage better. It is also vital to force grid operators to make it easier for smaller, more local forms of power generation to join the network and feed electricity into the grid. But ultimately, the responsibility is ours: we have to educate ourselves about the price of power and realize that there is nothing natural about the monopoly that the electric power companies — and, by extension, Big Coal — have over our lives. The system we have today is the one-hundred-year legacy of Thomas Edison and Samuel Insull. It has evolved into a frighteningly complex and powerful entity, but it was built with human ingenuity — and can be modified and disassembled with human ingenuity.

Finally, we need to grasp the urgent fact that every ton of CO_2 that a big coal burner pumps into the air pushes our world — and our children's world — that much closer to potential catastrophe. No one is arguing that coal plants should be shut down tomorrow, but we need to acknowledge that global warming is real and that we need to begin preparing for what's ahead. In the broadest terms, that means changing everything from how we subsidize coastal development to how we plan for natural disasters and floods of environmental refugees. But it also means finding a way to cut greenhouse gas emissions. We can argue about the best way to do this — ratify the Kyoto treaty, tax CO_2 emissions, expand the regional cap-and-trade program that has been proposed by northeastern governors, provide bigger subsidies to clean power development, or give away free bicycles. But after Hurricane Katrina, which laid bare just how unprepared we are for climate chaos, "the time to debate whether or not to act is over," according to Alex Steffen, the editor of WorldChanging, an influential energy and technology Web site. Steffen wrote:

> That debate's history, like the Berlin Wall. Katrina flattened it. In the aftermath of Katrina, we can no longer tolerate self-interest masked as caution, short-sightedness masked as responsibility, and lies masked as patriotism. To see the pictures and hear the stories coming out of New Orleans is to know one thing: whatever moral credibility professional environmental "skeptics" once claimed is as shredded as the Superdome roof.
>
> We aren't scrambling to reinvent our industrial civilization because we're bored. We aren't working for a more just global economy for kicks. We aren't fighting for democracy and human rights and good global governance in order to have something to talk about at parties. We aren't ringing the alarm sirens over global warming because we like the way they sound.
>
> We're doing all these things because the future of our planet is at stake. People's lives are at stake, millions of them.

Right now, the biggest impediment to change — and one that Big Coal does its best to encourage — is the idea that we are dependent on the very thing that is killing us. We hear it all the time: if we pass laws that limit CO_2 emissions, the price of electricity will skyrocket and the economy will collapse! If we clean up dirty coal plants, the price of electricity will skyrocket and the economy will collapse! If we restrict mountaintop removal mining, the price of electricity will skyrocket and the economy will collapse! These arguments amount to nothing less than blackmail.

This is nothing new in America. "The arguments against slavery were always bumping up against this: 'But it's an institution that's been around forever! What would happen if we got rid of it? How would you pay the people who lost their slaves, their valuable property? How would we harvest? It's not practical. What would we do?'" the writer Ian Frazier observed. "Lincoln's great moment was saying, 'I don't care if it's destructive. Slavery is wrong.' You start with, 'Is it right or wrong?' Then you act on that judgment. You don't say, 'I'm not going to say it's wrong because it would be too impractical to undo.'"

Burning coal may not be the moral equivalent of slavery, but it is a moral question nevertheless. In this context, it's important to acknowledge that Big Coal may in fact be right when it argues that reforming the industry will cause electricity prices to rise and the consumption of coal to fall. This could very well bring real economic hardship to some areas of the country, especially in regions where coal is mined. I grew up in a working-class family myself and know very well how devastating it can be to be caught on the wrong side of broad technological shifts. Even if those predictions turn out to be accurate, however, it still doesn't justify the larger costs and consequences of our continued indulgence in cheap coal.

But the truth is, I believe the industry grossly exaggerates the hardships that await us if we dare to change our thinking. Maybe it's just my entrepreneurial bias, but spending three years researching this book led me to feel in some ways *more* hopeful about the future, not less. I understand now that while the challenges the world faces in figuring out how to satisfy its thirst for energy without destroying itself in the process are immense, so are the opportunities. Old coal plants are more than just relics of an earlier era; they are giant bulwarks against change, mechanical beasts that are holding back a flood of ideas and innovation. When we muster up the courage to knock them down, the revolution will begin. It's not that I have blind faith that technology will save us or that I think we can snap our fingers and replace all the coal plants in the world with wind turbines and solar panels; I simply believe that it's within our grasp to figure out less destructive ways to create and consume the energy we need. Ultimately, the most valuable fuel for the future is not coal or oil, but imagination and ingenuity. We have reinvented our world before. Why can't we do it again? As a bumper sticker I saw in Wyoming put it, "Please, Lord, give me one more boom and I promise not to screw it up."

Afterword

While I was writing *Big Coal*, I sometimes feared the book would be obsolete before it arrived in bookstores. This is always a risk when you're writing about current events, but it is all the more dangerous when you're covering a subject as fast-moving and complex as energy. A single headline — MYSTERY OF COLD FUSION SOLVED BY NEW JERSEY RESEARCHER — could change everything.

Of course, no such breakthroughs occurred. In the United States, we consumed over a billion tons of black rocks in 2006, just slightly less than in 2005 (the 1 percent decline is mostly due to weather variations and railroad delivery problems). To put it another way, despite all the media hoopla about green energy, despite *An Inconvenient Truth* and the millions of compact fluorescent light bulbs sold by Wal-Mart and Home Depot, little has changed: the mountains of Appalachia are still being blown up, coal miners are still dying in the darkness, and the planet is heating up faster than ever. If anything, the events of the past year have only underscored the urgency of coming to grips with our addiction to coal.

The central issue remains the same today as it was when I finished this book: global warming. Is it possible to continue burning coal without cooking the climate? NASA's James Hansen, one of the top climatologists in the world, recently stated that the earth is very close to the hottest it has been in a million years, and that we have until 2015 to reverse the flow of carbon into the atmosphere before we cross a threshold and create a "different planet."

No worries, Big Coal tells us. In 2006 — very much like in 2005, and 2004, and 2003 — industry boosters suggested that technological breakthroughs were just over the horizon. Besides the usual talk about FutureGen, many press releases touted investments in technology that might someday remove CO_2 from traditional power plant stacks — perhaps even by using algae, which could then be harvested, dried, and burned in the plant as a biofuel. Who knows?

Right now, however, the coal industry is still playing the same old song. Of the 150 or so new coal plants planned in the United States today, all but a handful use traditional combustion coal technology. More important, none of these combustion plants include any serious plans to capture CO_2 when they are built. In China, a new coal plant is rising every ten days. This mad romance with coal is already having a measurable impact on the atmosphere. In 2000, CO_2 emissions were rising less than 1 percent annually. Today, they are rising more than 2.5 percent annually, with 7.9 billion metric tons of carbon added globally in 2005 alone (up from 6.8 billion in 2000). The Energy Department's latest report projects that the United States' CO_2 emissions will rise by one third from 2005 to 2030. In China and India, emissions will increase even more quickly.

But changes indeed loom on the horizon. When I began this book in 2002, it was still possible to find respected scientists who had doubts about global warming. By 2006, however, the accumulated weight of evidence in favor of human-induced warming was so great that principled skeptics were nowhere to be found. Not that there aren't many aspects of the earth's climate that researchers still don't understand. But any scientist who seriously questions the notion that global warming is caused by human activity (i.e., burning fossil fuels) is about as trustworthy as a doctor who denies that smoking causes cancer.

Today, the question is not "Is it happening?" but "What are we going to do about it?" The latest science suggests we may not have a lot of time to study our options. Arctic sea ice coverage in March 2006 was the lowest in winter since measurements by satellite began in the early 1970s, and a team of NASA-funded scientists found that ice is melting so fast in the Arctic that the North Pole may be in the open sea in forty years. Another landmark study published in 2006 pointed to the risks of ocean acidification from rising CO_2 levels, which make it increasingly difficult for marine organisms to survive and endanger the ocean food chain.

More damaging to Big Coal, however, is the fading logic of its economic arguments against tackling global warming. As I chronicled in chapter 8, Big Coal has long maintained that mandatory restrictions on CO_2 emissions would cripple the U.S. economy. That logic took a severe hit in 2006, when the former World Bank economist and senior British government adviser Sir Nicholas Stern released a major report on the economic costs of global warming. According to Stern, global warming stands to shave up to 20 percent off the world's annual GDP by the end of the century. But if we act now, it will only cost 1 percent of the world's annual GDP to avoid the damage. "Whether or not Stern's numbers are ultimately accepted," David Roberts wrote in *Grist*, an online magazine, "he smashed once and for all the myth that our choice is between spending money fighting global warming and saving money doing nothing. It turns out doing nothing will cost far more."

Equally problematic for Big Coal is the emergence of powerful venture capitalists such as Vinod Khosla, a Silicon Valley pioneer who has thrown himself headlong into energy investments. When I visited him in his Palo Alto office in December 2006, Khosla had been running numbers on new coal plants versus large-scale solar thermal installations and had determined that, given coal's supply problems and future CO_2 costs, solar power was a far less risky investment than coal. "You have to be crazy to build a traditional coal plant today," Khosla told me.

The political landscape is changing, too. While I was researching this book, several coal company executives told me off the record that they believed the war in Iraq was the best thing that ever happened to the coal industry. After all, the mounting toll of dead soldiers and citizens in Iraq is a pretty effective argument for energy independence. And despite the fact that coal is a lousy substitute for oil (see chapter nine), energy independence is precisely the rationale that President Bush used in his 2007 State of the Union address to push for more "alternative" fuels, including liquefied coal. Billions of dollars worth of subsidies and tax breaks are sure to follow.

But however much the war in Iraq has helped to sell the idea of energy independence, it is also responsible for the Republican meltdown in the 2006 elections. The new Speaker of the House, Nancy Pelosi, a California Democrat with strong ties to Silicon Valley, has made tackling global warming a top priority. As I write this, no less than four legislative proposals to cap CO_2 emissions are circulating in Washington — which one will win out, and how effective it will end up being, is anyone's guess. But even within the coal industry, it is widely acknowledged that some sort of mandatory cap on CO_2 will be passed in the next few years. The dinosaur wing of the Big Coal alliance will undoubtedly battle these proposals every step of the way. Others, such as TXU, the big Dallas-based power company, are using the coming crackdown as an excuse to throw up a fleet of new coal plants in hopes that they will be grandfathered in under any coming regulations.

Not everyone is following the old road of delay and denial, however. Recently, some big coal burners, including Duke Energy and Florida Power & Light (which owns a piece of Plant Scherer, the coal plant I wrote about in chapter 7), joined with such corporate giants as General Electric and DuPont and environmental groups such as Environmental Defense and Natural Resources Defense Council to form a coalition called the U.S. Climate Action Partnership. The stated purpose of the group is to advocate legislation that will cut greenhouse gas emissions in the United States by 60 to 80 percent by 2050. "The science of global warming is clear," Jim Rogers, the chairman and CEO of Duke Energy, said at a press conference for the coalition in January 2007. "We know enough to act now. We must act now."

It sounds hopeful. It sounds bold. It sounds resolute. Unfortunately, Rogers said more or less the same thing to me when we talked in 2001. We didn't act then, and it remains to be seen whether or not we'll act now.

The world heats up, but Big Coal rumbles on.

Saratoga Springs, NY
January 2007

Acknowledgments

The idea for this book grew out of an assignment for the *New York Times Magazine.* I'm grateful to Daniel Zalewski for sending me down to West Virginia and encouraging me to look around, and to Adam Moss for having faith in what I came back with. Larry Gibson, Bill Raney, Mike Whitt, Larry Emerson, and John McDaniel introduced me to coal mining in the Mountain State. Susan Lapis generously gave me an aerial view of the mines. Laura Forman shared her love of West Virginia with me; her memory lingers in these pages.

In Wyoming, I'm indebted to Bob Greene, Travis Todd, Dick Turpin, Bret Clayton, Mark Davies, Penny and Joel Hjorth, Neil Dunbar, Ryan Williams, and Sam Western. Romeo Flores answered endless questions about the geology of the Powder River basin. Dustin Bleizeffer helped me make some important connections and made sure I didn't leave Gillette without wrestling a few cattle to the ground.

It was a privilege to know the nine Quecreek miners, especially Randy Fogle, Mark Popernack, and Blaine Mayhugh, who showed remarkable courage and strength during some very trying times (and I don't mean down in the mine). John Straka introduced me to life underground. Elwood Reid, David Frankel, and Larry Sanitsky made an impossible assignment feel like an adventure. Thanks to Howard Messer for walking me through the legal details and to my friend Mitch Gelman for throwing me into the middle of the action.

By far the most fun I had researching this book was riding across the prairie in the locomotive of a coal train. I'm grateful to Pat Hiatte and Will Cunningham at BNSF for letting me climb aboard. Mike Mulder, Terry

Fox, Mike Callan, Ralph Peterson, David Clyde, and Mike Dafney were all great company.

In Atlanta, I learned a lot from Amoi Jeter, John Sell, Danny Morton, Butch Copelan, Alan Reeves, Walt Abston, and others at the Southern Company who shall remain nameless; Michael Chang, Sally Bethea, Robert Ukeiley, Colleen Kiernan, Becky Stanfield, and Ogden Doremus were all helpful. Thanks most of all to Neill Herring, who not only has an encyclopedic knowledge of Georgia politics, but is one of the few environmental advocates I've met who also has a deep appreciation for the industrial history of America.

In China, Zhang Jianyu, Ma Zhong, Wang Hao, and Wang Yongshen helped me leap over many cultural barriers. Woods Hole Oceanographic Institution allowed me to spend a month in the North Atlantic aboard their research ship, the *R. V. Knorr*. Lloyd Keigwin, the chief scientist on the trip, spent hours talking with me about benthic forams and iceberg armadas. Keigwin is a model scientist, at once bold and ruthlessly honest, and I'm lucky to have had the chance to watch him work.

I also owe thanks to Ralph Hysong, Dale Simbeck, James Childress, Dr. Mark Frampton, Dr. George Thurston, Dr. Howard Frumpkin, Michael Skelly, Zach Willey, Chris Walker, Pat Daly, Peter Bradford, Tiffany Elliot, Ross Bannister, Tom Berry, Buddy May, John Thompson, John Walke, Armond Cohen, David Hawkins, Vivian Stockman, Carol Raben, Nicole St. Clair, Eric Etheridge, Devereux Carter, Lisa Graves Marcucci, Eric Nonacs, Pat Towers, and Adam Levy. And to my lunch partner in Saratoga, a man who spent years teaching me about power and politics — thank you for your generosity and, more important, your friendship.

My agent, Heather Schroder, is as tough as any coal miner; my editor, Anton Mueller, pushed me harder than he should have had to, and I'm grateful for it. His guidance and incisive editing gave this book shape and spirit. Manuscript editor Beth Fuller put up with my last-minute changes with great patience and good humor. Jann Wenner and Will Dana at *Rolling Stone* gave me the time off to write the book and the confidence that I had an important story to tell.

Michele, thank you for everything, including the stars and the moon. And to Milo, Georgia, and Grace, the three dynamos who power my future and who showed such patience with their absentee father: yes, Dad can come out of his office now and go for a bike ride.

Notes

Introduction

xii   *More than six thousand:* "How Many More Must Die in Perilous Mines?" *China Daily,* August 11, 2005.

355,000 people: Comparative Quantification of Health Risks (Geneva: World Health Organization, 2004), p. 1402, table 17.8.

our per capita consumption: This is my own calculation, which I made as follows: In 2003, China burned about 1.5 billion tons of coal (U.S. Department of Energy, "International Coal Consumption Information," fig. 1.4, www.eia.doe.gov/emeu/international/coal.html#intlconsumption), and China's population was 1.3 billion *(Wikipedia),* so China's coal consumption was 1.15 tons per person. That same year, the United States burned about 1.1 billion tons of coal ("International Coal Consumption Information," fig. 1.4), and the U.S. population was 298 million *(Wikipedia),* which works out to be about 3.7 tons of coal consumed per person.

The average American consumes: 3.7 tons per person per year (7,400 pounds), divided by 365 days, equals 20.27 pounds per person per day.

xiii   *About half the electricity:* U.S. Department of Energy, Energy Information Administration, "Net Generation by Energy Source," 2004 data, www.eia.doe.gov/cneaf/electricity/epm/table1_1.html. Total electricity generation (in thousand megawatt-hours): 3,953,407. Total generation from coal: 1,976,333.

we burn more than: According to the Energy Information Administration, 1,105.4 million short tons of coal were consumed in the United States in 2004. Ninety-two percent of that coal was burned by the electric power industry. U.S. Department of Energy, Energy Information Administration, *Annual Coal Report,* 2004, www.eia.doe.gov/cneaf/coal/page/acr/acr_sum.html.

with revenues of more than: Edison Electric Institute, *EEI 2005 Financial Review,* "Consolidated Income Statement: U.S. Shareholder-Owned Electric Utilities," chart, p. 9, www.eei.org/industry_issues/finance_and_accounting/finance/research_and_analysis/financial_review/index.htm.

xiv *Between 1950 and 2000:* Tim Dyson, "On Development, Demography and Climate Change: The End of the World As We Know It?" (paper prepared for the twenty-fifth Conference of the International Union for the Scientific Study of Population, July 18–23, 2005), table 1.

By 2030, the world's demand: International Energy Agency, *World Energy Outlook 2004,* p. 29.

In 2005, the world consumed: BP, *BP Statistical Review of World Energy June 2006,* "Oil Consumption," chart, p. 11, www.bp.com/liveassets/bp_internet/globalbp/globalbp_uk_english/publications/energy_reviews_2006/STAGING/local_assets/downloads/pdf/statistical_review_of_world_energy_full_report_2006.pdf.

xv *consumption of natural gas:* "Unnatural Gas Prices," *Wall Street Journal,* May 3, 2005.

About 85 percent: U.S. Department of Energy, Energy Information Administration, "U.S. Natural Gas Imports and Exports: 2004," special report, December 2005, www.eia.doe.gov/pub/oil_gas/natural_gas/feature_articles/2005/ngimpex/ngimpexp.pdf.

natural gas prices have tripled: "Unnatural Gas Prices."

20 percent of the world's oil: BP Statistical Review of World Energy 2005, p. 4.

xvi *25 percent of the world's recoverable:* U.S. Department of Energy, Energy Information Administration, *International Energy Annual 2003,* International Coal Reserves Data, www.eia.doe.gov/emeu/international/coal.html#Reserves.

western Europe has only: Department of Energy, *International Energy Annual 2003,* International Coal Reserves Data.

Alternative energy guru: Amory Lovins, "More Profit with Less Carbon," *Scientific American,* September 2005.

the energy wasted: Ibid.

coal plants are responsible: U.S. Department of Energy, Energy Information Administration, "Carbon Dioxide Emissions," *Emissions of Greenhouse Gases in the United States 2003,* www.eia.doe.gov/oiaf/1605/ggrpt/carbon.html.

xix *The task force's recommendations:* National Energy Policy Development Group, "National Energy Policy: Report of the National Energy Policy Development Group," May 2001.

As of 2005, more than: National Energy Technology Lab, "Tracking New Coal-Fired Power Plants" (presentation, September 29, 2006).

xx *Since 1900, more than:* U.S. Department of Labor, Mine Safety and Health Administration, "Coal Fatalities for 1900 Through 2004," www .msha.gov/stats/centurystats/coalstats.htm.

Black lung, a disease common: Solid long-term statistics on black lung deaths are difficult to find. My estimate is an extrapolation of data from the National Institute for Occupational Safety and Health (NIOSH), which charted black lung deaths between 1968 and 1994 (www.courier journal.com/dust/illo_lungdeaths.html). During that time, the average yearly fatalities from black lung were about 2,100, with a high of 2,870 in 1973 and a low of 1,478 in 1994; 2,100 deaths per year, times 100 years, equals 210,000 total deaths. Because the number of miners working in the early part of the 1900s was far higher than between 1968 and 1994 and the conditions were much worse, this is a conservative estimate. See also "Dust, Deception and Death," a five-part investigation by Gardiner Harris and Ralph Dunlop, published in the *Louisville* (Ky.) *Courier-Journal,* April 19–26, 1998, www.courier-journal.com/dust/in dex.html.

buried more than 1,200 miles: U.S. Environmental Protection Agency, "Mountaintop Mining in Appalachia: Final Programmatic Environmental Impact Statement," October 2005.

If and when fruit trees: Intergovernmental Panel on Climate Change, *Third Assessment Report,* 2001, fig. 7.5.

In the 1920s, when: National Mining Association, "Growth of the Bituminous Coal Mining Industry in the United States" (data sheet, 2005).

Today there are more florists: Total coal miners, 2003: 62,148 (National Mining Association data sheet). Total florists, 2004: 111,570 (U.S. Department of Labor, Bureau of Statistics, "National Industry-Specific Occupational Employment and Wage Estimates").

Over the past 150 years: West Virginia Coal Association, *Coal Facts 2004,* p. 6.

The lowest median household income: U.S. Census Bureau, "Median Income of Households by State: 1984 to 2003," www.census.gov/hhes/ income/histinc/h08a.html.

a literacy rate: Jim Balow, "The Coalfields: What Comes Next?" *Charleston* (W.Va.) *Gazette,* July 25, 2004.

xxi *a generation of young people:* "New Generation Hits the Road," *Huntington* (W.Va.) *Herald-Dispatch,* April 28, 2002.

U.S. auto manufacturers: George Will, "What Ails GM," *Washington Post,* May 1, 2005; "Ailing GM Looks to Scale Back Generous Health Benefits," *USA Today,* June 23, 2005.

xxii *"I grew up in Logan":* Jennifer Stock's Web site, www.kdpublish.com.

xxiv *72,000 people:* American Lung Association, *State of the Air 2005,* lungaction.org/reports/sota05_full.html.

xxv *"We're playing":* Wally Broecker, interview, *Morning Edition,* National Public Radio, May 12, 2004.

"Chicago asked": Henry Adams, *The Education of Henry Adams* (New York: American Heritage Library, 1971), p. 343.

1. The Saudi Arabia of Coal

4 *"Coal? Wyoming has enough":* Fenimore Chatterton, quoted in John Thilenius and Gary Glass, "Surface Coal Mining in Wyoming: Needs for Research and Management," *Journal of Range Management,* September 1974, p. 337.

42 billion tons: EIA, *Annual Coal Report,* 2004, table 15, www.eia.doe .gov/cneaf/coal/page/acr/table15.pdf.

about 40 percent: Ibid., table 6, www.eia.doe.gov/cneaf/coal/page/acr/ table6.pdf.

5 *In 2002, at Arch Coal's:* U.S. Department of Labor, Mine Safety and Health Administration, report of investigation, "Fatal Fall of Highwall Accident," July 25, 2002. The accident occurred on February 20, 2002; the victim's name was Allen E. Greger, age forty-nine.

A month earlier: "State Inspector Disputes Mines Claim About Worker Death," *Billings* (Mont.) *Gazette,* April 3, 2002.

6 *When Tennessee senator:* Lamar Alexander, speech, "Importance of the Senate Passing an Energy Bill," U.S. Senate, June 15, 2005, alexander.sen ate.gov/index.cfm?fuseaction=speeches.detail&speech_id=77&month= 6&year=2005.

"almost 500 years": Larry J. Martin and Brian D. Winters, "Application of Wisconsin Energy Corporation for a Declaratory Ruling Pursuant to Section PSC 100.13(4)," Public Service Commission of Wisconsin, November 29, 2000, pp. 22–23.

"essentially unlimited": Peter Huber, quoted in "Email Altercation," *Los Angeles Times,* April 24, 2005.

7 *a big, shiny pile:* Much of the confusion about how much coal America has left comes from a misunderstanding of coal reserve terminology. The EIA uses three categories to classify reserves: Estimated Recoverable Reserves at Producing Mines, Estimated Recoverable Reserves, and Demonstrated Reserve Base. Most references to "500 years of coal" in America refer to the Demonstrated Reserve Base, which the EIA's 2004 Annual Coal Report (table 15) estimates at 494 billion tons. But citing this number as evidence of America's great bounty is highly misleading because it refers to a rough estimate of all coal in the United States that is *potentially* mineable. (According to the EIA, this includes all beds of bitu-

minous coal and anthracite 28 or more inches thick and beds of sub-
bituminous coal 60 or more inches thick that occur at depths of up to
1,000 feet, as well as beds of lignite 60 or more inches thick that can be
surface mined.) In addition to the fact that these coal estimates are based,
in part, on geological studies that are nearly one hundred years old, the
Demonstrated Reserve Base also does *not* take into consideration coal
quality, coal market economics, environmental restrictions, access to railroads or barge transportation, or dozens of other factors that have a real-
world impact on coal usage. Estimated Recoverable Reserves, which the
EIA estimates at 267 billion tons, attempts to roughly calculate some of
these factors — although as Jim Luppens, the head of the coal assessment program at the U.S. Geological Survey, put it to me, "There's a lot of
goat pasture in those numbers, too."

8   *"You hear the industry":* Interview with Nick Fedorko, October 2004.
    "People in the industry": Interview with Marshall Miller, October 2004.
    Coal is a creation: For my understanding of coal formation and geology, I
    relied on a number of sources, beginning with Barbara Freese, *Coal: A
    Human History* (New York: Perseus, 2003), pp. 17–21. Also important
    was John McPhee's masterful *Annals of the Former World* (New York:
    Farrar, Straus, and Giroux, 1998), particularly his description of the Lara-
    mide revolution in Wyoming (pp. 312–13). I also consulted R. M. Flores
    and L. R. Bader, "Fort Union Coal in the Powder River Basin, Wyoming,
    and Montana: A Synthesis," in U.S. Geological Survey Professional Paper
    1625-A, 1999.

11 *coal was put to use:* Freese, *Coal*, p. 112. See also "Mystery Surrounds 'Por-
    celain of the Southwest,'" *Science Daily*, September 7, 2005.
    In 1673, the French explorer: Pennsylvania Historical and Museum Com-
    mission, *A Walk Through the Rise and Fall of Anthracite*, p. 2, www.phmc
    .state.pa.us/ppet/miningmuseum/page2.asp?secid=31.
    In the 1750s: Freese, *Coal*, pp. 110–11.

12 *In 1874, a newspaper:* John F. Finerty, *War Path and Bivouac: The Big
    Horn and Yellowstone Expedition* (Chicago: R. R. Donnelley & Sons,
    1955), p. 244.
    the United States had a little over: Paul Averitt, *Coal Resources of the
    United States, January 1, 1974*, U.S. Geological Survey Bulletin 1412
    (Washington, D.C.: U.S. Department of the Interior, 1975), p. 49.

13 *the amount of coal:* Ibid., p. 33, table 5.
    only about 50 percent: Ibid., p. 32.
    In 1986, the USGS: The background of this project is described in the in-
    troduction to Timothy J. Rohrbacher et al., *Coal Reserves of the Matewan
    Quadrangle, Kentucky: A Coal Recoverability Study*, U.S. Bureau of
    Mines Information Circular, 1993.
    of the 986 million tons: Ibid., p. 10, figure 4.

Over the next couple: Devereux Carter et al., *Coal Availability, Recoverability, and Economic Evaluations of Coal Resources in the Northern and Central Appalachian Basin Coal Ranges,* U.S. Geological Survey Professional Paper 1625-C, 2000.

14 *at $25 a ton:* Rohrbacher, *Coal Reserves,* p. 7.

"If similar results": Ibid., p. 11.

In later studies: Timothy J. Rohrbacher et al., *Coal Availability, Recoverability, and Economic Evaluations of Coal Reserves in the Colorado Plateau,* U.S. Geological Survey Professional Paper 1625-B, 2000.

Even in the mighty: Lee M. Osmonson et al., *Coal Recoverability in the Hilight Quadrangle, Powder River Basin, Wyoming,* U.S. Geological Survey Open-File Report 00-103, February 2000.

"but statisticians are like": Interview with Timothy Rohrbacher, October 2004.

15 *"I'd like to be able":* Interview with Rich Bonskowki, October 2004.

The EIA claims: Department of Energy, *Annual Coal Report,* 2004, table 15; West Virginia Coal Association, *Coal Facts 2004,* p. 6.

16 *Coal mines are responsible:* U.S. Geological Survey, *Minerals Yearbook,* 2004, vol. I, pp. 24.1–24.2.

Blasting engineers: This story, first told to me by Gary Jerke at Cordero Rojo, may or may not be apocryphal. I tried to verify it with other blasters in the basin; they all had heard the story, but no one could confirm it. If it's a myth, it's a good one.

18 *Just ask the parents:* "Locals Protest Dangerous Coal Mining Practices," *Bristol* (Va.) *Herald Courier,* September 26, 2004. See also "Virginia Strip-Mining Death Brings Reforms," *All Things Considered,* National Public Radio, February 7, 2005.

19 *In Pennsylvania, miners now:* Frances X. Clines, "Mining Deep Underground Stirs Protest Above," *New York Times,* May 4, 2001.

In eastern Ohio: see "Commission Rules Against Ancient Ohio Forest: Coal Mining Allowed Under and Adjacent to Dysart Woods," *Free Press* (Columbus, Ohio), July 20, 2005. See also Buckeye Forest Council Web site, www.buckeyeforestcouncil.org/dysart/.

In Pike County, Kentucky: "Lick Creek Residents Try to Keep Coal Mine Out of Neighborhood," *Louisville Courier-Journal,* August 31, 2002.

In West Virginia, overloaded: "Coal's Victims: Just the Cost of Doing Business?" *Huntington Herald-Dispatch,* July 19, 2003.

2. Coal Colonies

21 *Early on a cold:* The details of the Massey Christmas Extravaganza came from interviews with people who attended the event, local journalists who covered it, and photographs and videotapes of the event. I attempted

to interview Claire Vaught, a member of the Massey Energy Spousal Group that organized the event, but she referred me to Massey's corporate headquarters. Massey's media relations department failed to return repeated phone calls and e-mail messages.

Southern West Virginia: Boone County produced 29,674,029 tons of coal in 2003, the most of any county in the state. See West Virginia Coal Association, *Coal Facts 2004,* p. 8.

the biggest coal company: Ibid., p. 4.

the $3.1 billion company: Calculated from figures on the Charles Schwab Web site, investing.schwab.com/trading/center?pwdmsg=msg1, November 4, 2005.

22 *"This is our way":* Personal communication with West Virginia reporter, December 2004.

His compensation in 2004: "Executive Compensation," *Richmond* (Va.) *Times-Dispatch,* June 12, 2005. Blankenship's compensation for 2004 totaled $6,442,898, the highest of any executive in West Virginia.

cashed in Massey stock: "Miners' Protections Gone," *Charleston Gazette,* November 23, 2004.

the company had lost: "Blankenship Used to Controversy," *Charleston Gazette,* November 11, 2004.

23 *James H. "Buck" Harless:* "Bush's Biggest W. Va. Backer," *Huntington Herald-Dispatch,* October 25, 2004. See also Michael Weisskopf and Adam Zagorin, "Cheney Gets Coal-Fired," *Time,* May 1, 2002.

Blankenship had spent more than: "Blankenship Used to Controversy."

A few months later: "Blankenship Lawsuit Called Publicity Stunt," *Charleston Gazette,* September 14, 2005.

Blankenship promptly sued: Ibid.

"the ultimate twenty-first-century": Chris Stirewalt, "In Politics, Don't Mess with Don," *State Journal* (Charleston, W.Va.), June 23, 2005.

"I'm a poor guy": Don Blankenship, quoted in "Blankenship Calls Pension Bonds a 'Crap Shoot,'" *State Journal,* June 8, 2005.

Blankenship, who often: "Blankenship Used to Controversy."

Blankenship's mother: Many details about Blankenship's upbringing can be found in "Blankenship Tells What Makes Him Tick," *Charleston* (W.Va.) *Daily Mail,* July 11, 2005; and "Captain of Industry," *Logan* (W.Va.) *Banner,* January 1, 2003. Blankenship's family history was confirmed for me by Barry Blankenship, Don's half brother.

24 *"I knew I had":* Don Blankenship, quoted in "Blankenship Tells What Makes Him Tick."

25 *In the spring of 2003:* My account of the floods was informed by numerous interviews with Maria Gunnoe, as well as by my visit to her home in West Virginia. The floods of 2003 were widely reported in local and national media, including "Saturated West Virginia Braces for More

Rain," *USA Today*, June 17, 2003; and "Mudslide Makes Creek a Monster," *Charleston Gazette*, June 16, 2003. A number of Web sites include dramatic video of the floods, including wvlightning.com/june162003 .shtml.

26 *Between 1995 and 2000:* Brian Bowling, "Massey Leads in Coal Permit Violations," *Charleston Daily Mail*, September 10, 2001.

The Martin County slurry: An excellent account of the Martin County spill, as well as the federal Mine Safety and Health Administration's blatantly corrupt investigation that followed, can be found in Clara Bingham, "Under Mined," *Washington Monthly*, January/February 2005. Although a criminal investigation was launched to look into allegations that Massey had falsified maps and other documents relating to the safety of the impoundment, no charges were filed. In 2002, Massey agreed to pay $3.25 million in fines and damages to the State of Kentucky. In an SEC filing in 2005, Massey said that it had spent about $77.9 million on cleanup and other spill-related costs, including civil claims, fines, and "other items." Of that, $74.3 million was paid for by insurance. See "Feds End Probe of Slurry Spill," *Charleston Gazette*, May 4, 2005.

The Environmental Protection Agency: Bingham, "Under Mined."

27 *"the nation's worst fatality record":* Ken Ward Jr., "Massey Safety Record Has Long Been an Issue," *Charleston Gazette*, January 21, 2006.

"Massey Energy has for several years": Quoted in ibid.

28 *"That such a country":* Joseph Diss Debar, quoted in John Alexander Williams, *West Virginia and the Captains of Industry* (Morgantown: West Virginia University Press, 2003), p. 166.

McDowell County lost: Jim Balow, "The Coalfields."

The region has some: Ibid.

"The economic conditions": Interview with Michael Hicks, November 2004.

29 *The best evidence:* Details about Edward Berwind and his coal empire are from Mildred Allen Beik, *The Miners of Windber* (University Park: Pennsylvania State University Press, 1996), pp. 9–11. See also *The Elms, 1901–2001* (Newport, R.I.: Preservation Society of Newport County, 2001), and *Newport Mansions: The Gilded Age* (Newport, R.I.: Preservation Society of Newport County, 1996).

30 *"worse than the conditions":* Quoted in Beik, *The Miners of Windber*, p. 299.

In 1906, three miners: Ibid., p. 212.

"dedicated to long-term": Berwind Corporation, "Fact Sheet," www.berwind.com. I made several attempts to reach executives at the Berwind Corporation to discuss the family's legacy but was unsuccessful.

31 *investments in real estate:* Ibid.

"There is something about": George Leader, quoted in Robert P. Wolensky et al., *The Knox Mine Disaster: The Final Years of the Northern Anthra-*

cite Industry and the Effort to Rebuild a Regional Economy (Pittsburgh: Pennsylvania Historical and Museum Commission, 1999), p. 128.

In 2005, the West Virginia: West Virginia Coal Association, *Coal Facts 2005*, p. 4.

Nigeria is a textbook example: P. Collier, "Aid, Shocks, and Trade: What East Timor Can Learn from African Experience," *East Timor: Development Challenges for the World's Newest Nation* (Singapore: Institute of Southeast Asian Studies, 2001).

32 *In a landmark 1995 study:* Jeffrey Sachs and Andrew Warner, *Natural Resource Abundance and Economic Growth*, Development Discussion Paper No. 517a, Harvard Institute for International Development, October 1995.

"All in all": Sheik Ahmed Yamani, quoted in Michael Ross, "The Political Economy of the Resource Curse," *World Politics* 51, no. 2 (1999), 297–322.

34 *"allowed men like Davis":* Williams, *West Virginia*, p. 174.

"a gray pall": Henry Caudill, *Night Comes to the Cumberlands* (Boston: Little, Brown, 1963), p. 325.

Coal-rich McDowell County: Balow, "The Coalfields."

35 *In West Virginia, four tons:* Department of Energy, *Annual Coal Report*, 2004, table 21, www.eia.gov/cneaf/coal/page/acr/table21.html.

Today the fund: Wyoming Legislative Service Office, Research memo, February 4, 2005.

36 *"That's venture capital":* Denise Giardina, from a brief essay she wrote during her failed run for the West Virginia governorship in 2000, takemehomewv.com/articles/elec-00-giardina.shtml (accessed January 2005).

recently revived: "McDowell Landfill Plan Resurrected," *Charleston Gazette*, February 17, 2005. One of the companies that was involved in the development of the dump, Capels Resources, is a subsidiary of Berwind Land.

38 *Seven people were killed:* National Oceanic and Atmospheric Administration, "Climate of 2003: Annual Review of Significant U.S. and Global Events," National Climatic Data Center, January 15, 2004, www.ncdc.noaa.gov/oa/climate/research/2003/ann/events.html.

In 2003, West Virginia: "State Leads Nation in FEMA Assistance," *Charleston Gazette*, July 23, 2004.

According to one EPA study: Alan Randall, *Estimating Environmental Damages from Surface Mining of Coal in Appalachia* (Cincinnati: Industrial Research Laboratory, U.S. Environmental Protection Agency, 1978), p. 67.

40 *objectionable "waste":* Joby Warrick, "Appalachia Paying the Price for White House Rule Change," *Washington Post*, August 17, 2004.

one federal study: Ibid.

the failure of: An account of this flood can be found in Gerald M. Stern, *The Buffalo Creek Disaster* (New York: Random House, 1976).

about 135 slurry impoundments: In recent years, West Virginia has developed an excellent coal slurry impoundment warning and monitoring program. For information, see the Coal Impoundment Location and Warning System Web site, www.coalimpoundment.org.

A few years ago: Interview with Dr. Diane Shafer, October 2004.

41 *Stout tested the water:* Ben M. Stout III, "Well Water Quality in the Vicinity of a Coal Slurry Impoundment Near Williamson, West Virginia" (prepared in response to questions from citizens attending the January 15, 2004, training session of the Coal Impoundment Location and Warning System, Delbarton, W.Va., December 10, 2004).

42 *"We are taking":* Interview with Ben Stout, September 2004.

Consider this TV commercial: I viewed the commercial, identified as "Parent and Teacher Meeting," on Massey's Web site, www.massey energyco.com, February 3, 2005.

44 *"If the likes":* Serge Halimi, "What's the Matter with West Virginia?" *Le Monde,* October 2004.

But an editorial: Dan Radmacher, "Arch Coal's Hypocrisy Is Maddening," Editorial, *Charleston Gazette,* April 9, 1999.

Blankenship's group was called: For more about Blankenship's fight to unseat McGraw, see "Political Ads Aired in D.C. Target W. Va. Audience," *Washington Post,* November 1, 2004; and "Benjamin Knocks Warren McGraw off Supreme Court," *Charleston Daily Mail,* November 3, 2004.

45 *In 2002, a Boone County:* "Massey's Political Donations Questioned," *Charleston Gazette,* October 21, 2004.

"I'm not bought": "Benjamin Knocks Warren McGraw off Supreme Court."

a shipping crate: "West Virginia Justice Expected to Be a Reliable Pro-Business Vote on Court," *Charleston Gazette,* January 11, 2005.

46 *"The poor showing":* "Money Isn't Enough," editorial, *Charleston Gazette,* November 9, 2006.

3. Dogholes

49 *"it's no longer":* Jeffrey Immelt, quoted in Amanda Griscom Little, "It Was Just My Ecomagination," *Grist,* May 10, 2005. The account of the GE ecomagination party is also drawn from Little. I viewed the "Model Miners" commercial on GE's Web site, ge.ecomagination.com.

"The commercial is sure": "GE Ads Generate Heat," *Chicago Sun-Times,* May 19, 2004.

"Even if coal": Seth Stevenson, "Coal Miners Hotter," *Slate,* May 31, 2005, www.slate.com/id/2119668.

52 *"what I call":* "President Says Quecreek Nine, Rescuers Typify Nation's Strength," *Pittsburgh Post-Gazette,* August 6, 2002.

$50,000 a year: National Mining Association, "Profile of the U.S. Coal Miner — 2004," www.nma.org/pdf/c_profile.pdf.

working in a coal mine: Coalition for Affordable and Reliable Energy Web site, www.careenergy.com/technology/health_safety.asp (accessed November 4, 2005).

53 *mining is still:* National Institute for Occupational Safety and Health, *Worker Health Chartbook 2004,* NIOSH Publication No. 2004-146, fig. 4.1. The top three fatal occupational injury rates per 100,000 workers: mining (all types, but the group is made up predominantly of coal miners) 23.5 deaths; agriculture, forestry, and fishing 22.7; and construction, 12.2.

Working in an underground: In underground coal mines run by independent contractors, NIOSH estimates a fatality rate of 86.5 per 100,000. In large corporate-owned strip mines, the death rate is 17.6 per 100,000. For commercial fishing in Alaska, NIOSH estimates a fatality rate of 124 per 100,000. See ibid.; and "Traumatic Occupational Injuries — Commercial Fishing in Alaska," NIOSH Web site, www.cdc.gov/niosh/injury/traumafish.html.

killed in a methane explosion: David Jackson, "The Human Cost of Coal Mining," *Chicago Tribune,* September 22–24, 2002.

more than 1,500: "Dust, Deception and Death."

According to U.S. Department of Labor: U.S. Department of Labor, Bureau of Labor Statistics, "Average Weekly Earnings of Production Workers: Coal Mining," NAICS Code 2121 (inflation calculated with BLS inflation calculator on Web site), data.bls.gov/PDQ/servlet/surveyoutputservlet (accessed November 16, 2005); see also U.S. Department of Energy, Energy Information Administration, *Annual Energy Review 2004,* "Coal Mining Productivity, Selected Years 1949–2004," table 7.6, p. 215.

union mines in the East: Jennifer B. Leinart, "U.S. Coal Mine Salaries, Wages, and Benefits: 2004 Survey Results," report, Western Mine Engineering, Inc. (Spokane, Wash.), pp. 8–9.

54 *Today the union represents:* Interview with Doug Gibson, United Mine Workers of America, December 2004.

Blaine Mayhugh, the youngest: Information about Mayhugh's life, views, and work history are taken from a series of interviews I conducted with him between August and October 2002.

55 *"Think of hell":* Edgar Allen Forbes, quoted in "Loss of Life Leads to Change," *Charleston Daily Mail,* February 9, 1999.

56 *Some historians believe:* Lockard, *Coal,* p. 64.

PBS was a subsidiary: Charles McCollester, "Less Than Miraculous," *Nation,* March 17, 2003.

58 *dozens of local residents:* "Clues Overlooked About Mine," *Pittsburgh Tri-bune-Review*, August 3, 2002.
"*Obviously the firm*": Jeffrey Bender, quoted in ibid.
Pennsylvania is honeycombed: Allison M. Heinrichs, "Reclaiming Old Mines Is a Race Against Time," *Pittsburgh Tribune-Review*, August 7, 2005; see also John J. Fialka, "Perilous Pits," *Wall Street Journal*, June 4, 2003.

59 *PBS Coals, along with:* Pennsylvania Department of Environmental Pro-tection, Bureau of Deep Mine Safety, "Report of Investigation: Non-In-jury Mine Inundation Accident (Entrapment), July 24, 2002," July 22, 2003, p. 17.

60 *the disaster in Monongah:* Lockard, *Coal*, pp. 65–66.
"*especially with respect*": Quoted in ibid., pp. 66–67.
In 1947, an explosion: Ibid., pp. 66–71.

61 "*If we must*": John L. Lewis, quoted in Maier B. Fox, *United We Stand: The United Mine Workers of America, 1890–1990* (n.p.: United Mine Workers of America, 1990).
In 1968, an explosion: Lockard, *Coal*, p. 71.
"*Let me assure you*": Stewart Udall, quoted in ibid., p. 71.
the rate of fatal accidents: Ibid., pp. 72–73.
"*the [Interior] Department's*": Quoted in ibid., p. 83.

62 "*an agency in trouble*": Orrin Hatch, quoted in ibid., p. 85.
"*culture change*": Dave Lauriski, remarks, Wyoming Mining Association Safety and Reclamation Luncheon (Moran, Wyoming), June 21, 2002.
from 1997 to 2000: Bingham, "Under Mined."
Chao even hired: Ibid.

63 "*Mitch McConnell calls me*": Bob Murray, quoted in ibid.
On September 23, 2001: This account of the explosion at the Jim Walter Resources No. 5 mine and its aftermath is drawn from David Jackson's excellent three-part series "The Human Cost of Coal Mining."

64 "*If you know*": Joe Main, quoted in Adam Roston, "Fire in the Hole," *Mother Jones*, September/October 2002.
"*Where is MSHA?*": Senator Robert Byrd, quoted in Ken Ward Jr., "MSHA Goes to Court for Union," *Charleston Gazette*, January 26, 2006.
In a letter to Labor Secretary Chao: U.S. House of Representatives, Com-mittee on Education and the Workforce, "Lawmakers Say Bush Adminis-tration Has Always Had Authority to Issue Large Mine Safety Fines, But Has Refused to Use It," press release, January 25, 2006. http://www.house.gov/apps/list/press/ed31_democrats/rel12506.html.

66 *The average age:* National Mining Association, "Profile."

67 "*They've taken*": Interview with Christopher Bise, February 2004.
The breakthrough: This account is drawn from interviews I conducted with the Quecreek miners in August and September 2002. A more de-

tailed account can be found in their book, as told to me, *Our Story: Seventy-Seven Hours That Tested Our Friendship and Our Faith* (New York: Hyperion, 2002).

69 *On January 22, 1959:* Robert P. Wolensky et al., *The Knox Mine Disaster* (Harrisburg: Pennsylvania Historical and Museum Commission, 1999).

70 *The MSHA report:* U.S. Department of Labor, Mine Safety, and Health Administration, "Report of Investigation: Underground Coal Mine, Non-Fatal Entrapment, July 24, 2002," Quecreek #1 Mine, ID #36-08746, August 12, 2003, pp. 68–69.

71 *Total fines:* "Quecreek Accident Results in Fines," *Pittsburgh Tribune-Review,* June 9, 2004.
   In a report to Kathleen McGinty: Harry Hart, senior deputy inspector general for the State of Pennsylvania, to Kathleen McGinty, "Re: Section 236, Bituminous Coal Mine Act," April 25, 2003.
   "jumping on": Ibid., p. 6.
   the first appeared: "Quecreek Review Blames Bad Maps," *Pittsburgh Tribune-Review,* November 8, 2002.
   the second was found: "Forgotten Maps Could Have Averted Mine Flood," *Pittsburgh Tribune-Review,* January 15, 2005.

72 *Pennsylvania taxpayers:* "Quecreek Costs Near $2 Million," *Pittsburgh Tribune-Review,* January 4, 2003.
   In early 2005: "Transformer Explosion Injures Two Quecreek Miners," *Philadelphia Inquirer,* January 15, 2005.
   Meanwhile, MSHA chief: "Kerry Calls for Federal Probe of MSHA Contracts," *Charleston Gazette,* November 2, 2004.

4. The Carbon Express

75 *Canals were the first:* An excellent account of the early days of coal-hauling canals and railroads can be found in Donald L. Miller and Richard E. Sharpless, *The Kingdom of Coal: Work, Enterprise, and Ethnic Communities in the Mine Fields* (Easton, Pa.: Canal History and Technology Press, 1998), pp. 19–51. See also Freese, *Coal,* pp. 90–95.
   250 million tons: Burlington Northern Sante Fe, *2005 Annual Report,* p. 2.

77 *More than 20 percent:* Matt Rose, "Managing Prosperity in Coal," presentation, National Coal Transportation Association meeting, September 13, 2005.

78 *BNSF was a supporting:* Paul R. Samson, "Non-State Actors and Environmental Assessment: North American Acid Rain and Global Climate Change," ENRP Discussion Paper E-98-10, Kennedy School of Government, Harvard University, 1998, p. 34.

until recently: Ned Leonard, executive director of the Greening Earth Society, to David Appell, journalist, nasw.org/users/appell/Weblog/indexold.html.

"I think the future": Matt Rose, quoted in "Alliance Welcomes Rose at Town Hall," *Alliance Mechanical* (BNSF newsletter), September 2003, p. 4.

79 *Arch Coal, which owns:* "Freight Shipping Lags amid Economic Surge," *Washington Times,* July 13, 2004.

80 *"If there's one thing":* Interview with Bret Clayton, June 2003.

81 *The route was laid out:* Edward Gillette, *Locating the Iron Trail* (Boston: Christopher Publishing House, 1925), pp. 57–90.

83 *In 2005, after coal:* "A Mountain of Coal Waits for a Ride," *USA Today,* August 25, 2005.

84 *"The only control":* Interview with Greg Henshaw, February 2004.

85 *One example is:* See Terry Huval, testimony before House Transportation and Infrastructure Committee, Subcommittee on Railroads, March 31, 2004.

50 percent higher: Ibid.

86 *"They know how":* Henshaw, interview.

87 *BNSF secretly ran:* "Genetic Testing," *The NewsHour with Jim Lehrer,* PBS, June 7, 2001.

89 *It happened at 3:30 A.M.:* National Transportation Board, "Railroad Accident Report: Collision and Derailment Involving Three Burlington Northern Freight Trains Near Thedford, Nebraska, June 8, 1994," September 7, 1995.

90 *As the* Chicago Tribune: "Overworking on the Railroad," *Chicago Tribune,* July 13, 2004.

91 *"I then called":* E-mail message from Bob Kerrey, August 20, 2005.

5. Infinite Needs

99 *"The symptoms include":* American Psychiatric Association, *Diagnostic and Statistical Manual of Mental Disorders,* 4th ed. (Washington, D.C.: American Psychiatric Association, 2000), pp. 175–81.

Between 1970 and 2000: National Mining Association, "Fast Facts About Coal," www.nma.org/statistics/pub_fast_facts.asp.

In the 1920s, for example: Wade H. Wright, *History of the Georgia Power Company, 1855–1956* (Atlanta: Georgia Power, 1957), p. 334.

Today Georgia Power: Southern Company, "Georgia Power Facts and Figures," www.southernco.com/gapower/about/facts.asp.

101 *"descended from artisans":* Forrest McDonald, *Insull* (Chicago: University of Chicago Press, 1962), p. 4.

"Reading about": Samuel Insull, *The Memoirs of Samuel Insull* (Polo, Ill.: Transportation Trails, 1992), p. 19.

102 *"Long-Legged Mary Ann":* Jill Jonnes, *Empires of Light: Edison, Tesla, Westinghouse, and the Race to Electrify the World* (New York: Random House, 2003), p. 61. See also Mark Essig, *Edison and the Electric Chair: A Story of Light and Death* (New York: Walker and Company, 2003), p. 71.

103 *"more than 50,000 years":* Thomas Edison, quoted in Margaret Cheney, *Tesla: Man Out of Time* (New York: Touchstone, 2001), p. 163.
Instead of creating: Mark Granovetter and Patrick McGuire, "The Making of an Industry: Electricity in the United States," in *The Laws of Markets,* ed. Michel Callon (Oxford: Blackwell, 1998), pp. 147–73. See also Jean Strouse, *Morgan: American Financier* (New York: Random House, 1999), p. 313.

104 *saw it as a mysterious:* Linda Simon, *Dark Light: Electricity and Anxiety from the Telegraph to the X-Ray* (New York: Harcourt, 2004), pp. 242–45.
"Do it big, Sammy": Thomas Edison, quoted in McDonald, *Insull,* p. 38.

105 *"economically wrong":* Samuel Insull, quoted in Richard Rudolph and Scott Ridley, *Power Struggle: The Hundred-Year War over Electricity* (New York: Harper and Row, 1986), p. 38.
"natural monopolies": Ibid.
"No shrewder piece": Daniel Hoan, quoted in ibid., p. 40.

106 *if prices were low enough:* McDonald, *Insull,* pp. 64–69; see also Richard Munson, *From Edison to Enron: The Business of Power and What It Means for the Future of Electricity* (Westport, Conn.: Praeger, 2005), p. 48.
As early as 1914: Adrian Forty, *Objects of Desire: Design and Society from Wedgwood to IBM* (London: Thames and Hudson, 1986), pp. 186–87.

107 *"It was not until":* David E. Nye, *Consuming Power: A Social History of American Energies* (Boston: MIT Press, 1999), p. 171.
"For Health's Sake": Forty, *Objects of Desire,* p. 191.
"How long should": Harold L. Platt, *The Electric City: Energy and Growth of the Chicago Area, 1880–1930* (Chicago: University of Chicago Press, 1991), p. 237.
a pioneer in advertising to children: Munson, *From Edison to Enron,* p. 70.

108 *"massing production":* McDonald, *Insull,* p. 98.

109 *Between the 1920s:* Nye, *Consuming Power,* p. 198.
sixty times larger: Munson, *From Edison to Enron,* p. 54.
By 1916, Insull controlled: Rudolph and Ridley, *Power Struggle,* p. 42.
By the 1920s, Insull: Ibid., pp. 55–56.
In 1905, less than: Nye, *Consuming Power,* p. 171.

110 *a Yankee blue blood:* Unsigned biographical sketch in Atkinson family file, November 5, 1927, Atlanta History Center. Harry Atkinson was born in 1862 in Brookline, Massachusetts. The Atkinsons were an old and prosperous New England family that included many prominent liber-

als. Harry's uncle Edward, a political essayist and reformer, had helped finance and arm abolitionist John Brown for his disastrous raid on a military arsenal at Harpers Ferry, Virginia (now West Virginia), in 1859. See Harold Francis Williamson, *Edward Atkinson: The Biography of an American Liberal* (Boston: Old Corner Book Store, 1934), p. 4.

Commonwealth & Southern eventually: "Officials Silent upon Power Ruling," *Atlanta Journal*, March 20, 1941.

at its peak: Platt, *The Electric City*, p. 272.

111 *"can no more contest":* George Norris, quoted in Rudolph and Ridley, *Power Struggle*, pp. 53–54.

"un-American" and tied to "Bolshevik": Ibid., p. 50.

"a systematic, subtle": Franklin Roosevelt, quoted in ibid., p. 52.

"The Insull failure": Franklin Roosevelt, quoted in ibid., p. 67.

112 *"the gospel of consumption":* Platt, *The Electric City*, p. 276.

113 *"the golden years":* Smithsonian Institution, "Post World War II 'Golden Years,'" americanhistory.si.edu/powering/past/h2main.htm. This article is part of a project called *Powering a Generation of Change*, which documents the restructuring of the electric power industry.

rising by 60 percent: Vijay Vaitheeswarn, *Power to the People: How the Coming Energy Revolution Will Transform an Industry, Change Our Lives, and Maybe Even Save the Planet* (New York: Farrar, Straus, and Giroux, 2003), p. 31.

Efficiency-wise: Thomas R. Casten and Brennan Downes, "Critical Thinking About Energy: The Case for Decentralized Generation of Electricity," *Skeptical Inquirer*, January/February 2005, p. 25.

114 *"The industry had ossified":* Ibid., p. 26.

115 *Some old coal burners:* "Top Plants Supplement," *Power*, July/August 2004. According to *Power*, the cheapest coal plant in America is the Gerald Gentleman Station in Sutherland, Nebraska, which generates electricity for $7.95 per megawatt-hour.

116 *"state of the art":* Robert Williams, "IGCC: Next Step on the Path to Gasification-Based Energy from Coal," report to the National Commission on Energy Policy (Princeton Environmental Institute, Princeton University, November 2004).

would add an average: Ibid., p. 17, table 1.

117 *"What we need":* "The Experimental Economist," interview with Vernon Smith, *Reason Online*, October 9, 2002, reason.com/hod/fe.ml.smith.shtml.

118 *"In my business":* McDonald, *Insull*, p. 106.

6. The Big Dirty

119 *More than 2 million:* National Center for Health Statistics, "Deaths: Preliminary Data for 2004," www.cdc.gov/nchs/deaths.htm.

120 *"We thought":* Interview with Charlotte O'Rourke, March 2005.

121 *sued by five states:* "Pa., Four Other States Sue Allegheny Energy," *Pittsburgh Business Times,* June 28, 2005.

122 *the average emissions rate:* Americans for Balanced Energy Choices, "Emissions Reductions," www.balancedenergy.org/abec/index.cfm?cid=7540,7605.

   Nationwide, power plants: Ilan Levin and Eric Schaeffer, "Dirty Kilowatts: America's Most Polluting Power Plants" (report, Environmental Integrity Project, May 2005).

123 *Coal plants also release:* National Environmental Trust, "Beyond Mercury" (report, August 2004).

   Each year, coal plants produce: U.S. Environmental Protection Agency, "Notice of Regulatory Determination on Wastes from the Combustion of Fossil Fuels," May 22, 2000, p. 32216.

   half of all Americans: U.S. Environmental Protection Agency, "National Air Quality and Emissions Trends Report: 2003 Special Studies Edition," September 2003, p. 8.

   In the Northeast: U.S. Environmental Protection Agency, "The Particle Pollution Report," December 2004, p. 15, fig. 13.

   Levels of ozone: American Lung Association, *State of the Air: 2004,* lungaction.org/reports/sota_94_full.html.

   In 2004, 29 million children: Ibid.

   African Americans: Clean Air Task Force, *Dirty Air, Dirty Power: Mortality and Health Damage Due to Air Pollution from Power Plants,* June 2004, p. 10, www.catf.us/publications/view/24.

   per capita mortality: Ibid., p. 14.

   researchers have learned: National Research Council, *Research Priorities for Airborne Particulate Matter. IV: Continuing Research Progress* (Washington, D.C.: National Academies Press, 2004), p. 104. See also Jim Giles, "Nanoparticles in the Brain," *Nature,* June 9, 2004.

124 *children who grow up:* W. J. Gauderman et al., "Association Between Air Pollution and Lung Function Growth in Southern California Children," *American Journal of Respiratory and Critical Care Medicine* 166, no. 1 (2002): 76–84.

   "I'll bet you won't": Quoted in Margaret Newkirk, "Coal Loses Clout," *Akron (Ohio) Beacon Journal,* January 8, 2001.

   "death coming into": Quoted in Freese, *Coal,* p. 27.

125 *The life expectancy:* Ibid., p. 82.

   "Smoke is the incense": Quoted in J. R. McNeill, *Something New Under the Sun: An Environmental History of the Twenty-First Century* (New York: W. W. Norton, 2001), p. 59.

   "From a final cat-nap": Arnold Bennett, *Your United States: Impressions of a First Visit* (New York: Harper and Brothers, 1912), p. 110.

   "When I moved to Atlanta": Dr. Leila Denmark, quoted in Clifford M.

Kuhn, Harlon E. Joyce, and E. Bernard West, *Living Atlanta: An Oral History of the City, 1914–1948* (Atlanta: University of Georgia Press, 1990), p. 232.

126 *it was not until:* John R. Stilgoe, *Metropolitan Corridor: Railroads and the American Scene* (New Haven, Conn.: Yale University Press, 1983), p. 121.

The real awakening: For a full account of the Donora killer fog, see Devra Davis, *When Smoke Ran Like Water: Tales of Environmental Deception and the Battle Against Pollution* (New York: Basic Books, 2002), pp. 5–30.

127 *"America pays its debt":* Richard Nixon, quoted in ibid., p. 90.

"You better watch out": Ibid., p. 95.

128 *worked up a study:* Gary Stix, "Where the Bodies Lie," *Scientific American,* June 1998.

researchers now estimate: J. M. Samet et al., *The National Morbidity, Mortality, and Air Pollution Study. II: Morbidity and Mortality from Air Pollution in the United States,* The Health Effects Institute, 2000.

129 *He discovered that deaths:* Hillary Johnson, "The Next Battle over Clean Air," *Rolling Stone,* January 18, 2001.

130 *"a unique, natural experiment":* C. Arden Pope, quoted in ibid.

Among other things: D. W. Dockery et al., "An Association Between Air Pollution and Mortality in Six U.S. Cities," *New England Journal of Medicine* 329 (1993):1753–59.

"It was getting harder": Joel Schwartz, quoted in Johnson, "The Next Battle."

131 *"When I look":* Joel Schwartz, quoted in Stix, "Where the Bodies Lie."

Air Quality Standards Coalition: Marion Currinder, "Cultivating the Grass Roots," Capital Eye, Center for Responsive Politics, 1996, www.opensecrets.org/newsletter/ce44/ce44.02.htm.

Industry estimates ranged: Ibid.

"You look at the map": George Allen, quoted in Karl Blankenship, "New Rules Would Help Bay," *Bay Journal,* January/February 1997.

The EPA disagreed: Currinder, "Cultivating the Grass Roots."

"If we've learned": John McManus, quoted in Claudia H. Deutsch, "Still Defiant, but Subtler, Industry Awaits EPA Rules," *New York Times,* May 27, 1997.

"Spent a lot": Curtis Moore, "Lawsuit Against Clean Air Act by Members of Congress Raises Conflict-of-Interest Questions," Center for Public Integrity, September 5, 2000, www.publicintegrity.org/report.aspx?aid=392.

132 *"secret science":* Ibid.

"When you look": Interview with C. Arden Pope, October 2004.

"In some ways": Interview with Dr. Mark Frampton, September 2004.

133 *ultrafines show up:* G. Oberdörster et al., "Translocation of Inhaled Ultra-

fine Particles to the Brain," *Inhalation Toxicology* 16 (June 2004): 437–45.

estimates that about 14 percent: Interview with Dr. Lilian Calderon-Garciduenas, September 2004.

She found that: Lilian Calderon-Garciduenas et al., "Air Pollution and Brain Damage," *Toxicologic Pathology* 30, no. 3 (2002): 373–89.

"Air pollution doesn't": Calderon-Garciduenas, interview.

"It may turn out": Frampton, interview.

134 *In 1997, Dr. Karen Wetterhahn:* Karen Endicott, "The Trembling Edge of Science," *Dartmouth Alumni Magazine*, April 1998.

Coal-fired power plants: U.S. Environmental Protection Agency, "Frequent Questions About Mercury," www.epa.gov/mercury/faq.htm#14; U.S. Environmental Protection Agency, *Toxics Release Inventory*, 2003, www.epa.gov/triexplorer/chemical.htm. In addition to the 96,000 pounds of mercury that is released into the air by power plants, another 37,000 pounds is released into landfills and ash ponds. Many of these ash ponds and landfills meet federal requirements for the disposal of hazardous chemicals, but many others do not. In most cases, they were built before the disposal of hazardous chemicals was regulated.

135 *Some studies show:* Karen Wright, "Our Preferred Poison," *Discover*, March 2005.

In adults, recent studies: J. T. Salonen et al., "Intake of Mercury from Fish, Lipid Peroxidation, and the Risk of Myocardial Infarction and Coronary, Cardiovascular, and Any Death in Eastern Finnish Men," *Circulation* 91 (1995): 645.

Some researchers believe: Wright, "Our Preferred Poison."

one in twelve women: Centers for Disease Control, *Second National Report on Human Exposure to Environmental Chemicals*, 2003. See also Guy Gugliotta, "Mercury Threat to Fetus Raised, EPA Revises Risk Estimates," *Washington Post*, February 6, 2004.

136 *80 percent:* State and Territorial Air Pollution Program Administrators, *Regulating Mercury from Power Plants: A Model Rule for States and Localities*, November 2005, p. 14.

the deposition rate: Dr. David P. Krabbenhoft, "Mercury Emissions: State of the Science and Technology," statement before House Committee on Science, Subcommittee on Environment, Technology, and Standards, November 5, 2003.

the Greek historian Plutarch: Berton Roueche, *The Medical Detectives*, vol. 2 (New York: E. P. Dutton, 1984), p. 78.

"One year": Jonathan Watts, "Mercury Poisoning of Thousands Confirmed," *Guardian* (London), October 16, 2001.

137 *the Japanese government:* Akio Mishima, *The Bitter Sea: The Human Cost of Minamata Disease* (Tokyo: Kosei Publishing, 1992), pp. 129–40.

"The greater": Masazumi Harada, "Minamata Disease: A Medical Re-

port," in *Minamata*, W. Eugene Smith and Aileen Smith, eds. (New York: Holt, Rinehart, and Winston, 1975), pp. 180–91.

138 *EPA scientists estimate:* Kathryn Mahaffey et al., "Blood Organic Mercury and Dietary Mercury Intake: National Health and Nutrition Examination Survey, 1999 and 2000," *Environmental Health Perspectives* 112, no. 5 (April 2004): pp. 562–70.

"mercury emissions": U.S. Environmental Protection Agency, "EPA Decides Mercury Emissions from Power Plants Must Be Reduced" (press release, December 14, 2000).

"If we cannot": Quin Shea, "The Mercury Message," *Electric Perspectives*, September/October 2000.

"Mercury is a killer": Quin Shea, speech, Western Coal Transport Group, April 2001.

139 *Clear Skies:* "President Bush Announces Clear Skies and Global Climate Change Initiatives," White House fact sheet, February 14, 2002, www.whitehouse.gov/news/releases/2002/02/20020214.html.

23,000 people: U.S. Environmental Protection Agency, "Benefits and Costs of the Clean Air Act 1990–2010" (report to Congress, November 1999), p. 60, www.epa.gov/air/sect812/index.html.

141 *it's worth noting:* Chris Mooney, *The Republican War on Science* (New York: Basic Books, 2005), p. 137.

"At least eight": Deborah Rice, quoted in ibid.

142 *"We have learned":* Philippe Grandjean, quoted in Wright, "Our Preferred Poison."

"greatest hoax": Senator James Inhofe, "The Science of Climate Change," U.S. Senate floor statement, July 28, 2003, inhofe.senate.gov/press releases/climate.htm.

"The push to regulate": U.S. Chamber of Commerce, "Mercury Emissions," *Radio Actualities*, 2003, www.uschamber.com/issues/index/envi ronment/mercuryemissions.htm.

One coal industry expert: Larry S. Monroe, testimony, U.S. Senate Environment and Public Works Committee, June 5, 2003, epw.senate.gov/108th/Monroe_060503.htm.

143 *15 to 60 cents a month:* State and Territorial Air Pollution Program Administrators, "Regulating Mercury from Power Plants," p. 26.

144 *one EPA researcher:* Personal communication.

"EPA, in its expert": U.S. Environmental Protection Agency, "Revision of December 2000 Regulatory Finding on the Emissions of Hazardous Air Pollutants from Electric Utility Steam Generating Units and the Removal of Coal- and Oil-Fired Electric Utility Steam Generating Units from Section 112(c) List," March 29, 2005.

the Washington Post reported: Eric Panin, "Proposed Mercury Rules Bear Industry Mark," *Washington Post*, January 31, 2004.

net economic benefits: U.S. Government Accountability Office, "Observations on EPA's Cost-Benefit Analysis of Its Mercury Control Options" (report to congressional requesters, February 2005).

145 *"They treated us":* Interview with Dr. Randy Manning, December 2005.

146 *Between 1975 and 2001:* National Environmental Trust, "Beyond Mercury."

Toxic air emissions: According to the EPA's Toxics Release Inventory, 2003, electric utilities emitted 719,462,689 pounds of toxic chemicals into the air. Compare that with the chemical industry, which had the second-highest emissions, 168,993,305 pounds.

Human skeletal lead burdens: Jesse Ausubel, "Reasons to Worry About the Human Environment," *Cosmos* 8 (1998): 1–12.

dyslexia, attention deficit hyperactivity disorder: Ted Schettler et al., "In Harm's Way" (report, Greater Boston Physicians for Social Responsibility, May 2000).

"I worry": Ausubel, "Reasons to Worry."

7. *"A Citizen Wherever We Serve"*

148 *Southern's seventy-nine generating stations:* Southern Company, "Fact and Figures," www.southerncompany.com/aboutus/figures.asp

149 *"We're pleased":* David Ratcliffe, speech, Southern Company Annual Shareholders Meeting, Atlanta, May 25, 2005.

"kneecap breakers": Personal communication.

Between 2001 and 2004: Center for Public Integrity, LobbyWatch, www.publicintegrity.org/lobby/top.aspx?act=topcompanies.

Southern and its affiliates: Center for Responsive Politics, "Electric Utilities: Top Contributors to Federal Candidates and Parties," www.open secrets.org/industries/contrib.asp?Ind=E08&Cycle=2004. The center notes that total contributions "are based on contributions from PACs, soft money donors, and individuals giving $200 or more."

The Washington Post *reported:* Thomas B. Edsall, "Republican Energy Bill Gives Away Billions to Investors," *Washington Post*, November 22, 2003.

150 *For 2004, this group:* Public Citizen database, www.whitehouseforsale.org.

151 *"the* T. rex *of the power industry":* Interview with Joe Tondu, November 2005.

"what drives some people": Personal communication.

"a trench warfare mentality": Personal communication.

The Georgia Power Foundation: Cindy S. Theiler, "Georgia Power Foundation," *The New Georgia Encyclopedia*, www.georgiaencyclopedia.org/nge/Article.jsp?id=h-3423.

153 *"Historically":* Rudolph and Ridley, *Power Struggle*, p. 184.

In 2001, Georgia Power: Margaret Newkirk, "Rate Case Down to the Wire," *Atlanta Journal-Constitution,* December 19, 2004.

Golden Sleaze Award: Scott Henry, "The Sixteenth Annual Golden Sleaze Awards," *Creative Loafing,* March 30, 2005.

154 *"It was no more a rate cut":* Newkirk, "Rate Case Down to the Wire."

156 *"It was a very scary time":* Bob Deans, "Scherer: '75 'Scary' for Ga. Power," *Atlanta Journal-Constitution,* May 20, 1984.

In 1989, federal agents: Gail Epstein and David Secrest, "Ga. Power Files Seized in Raid by IRS," *Atlanta Journal-Constitution,* July 22, 1989; Mark Whitaker, "File on 4," BBC, January 30, 1996; Greg Palast, "Enron: Not the Only Bad Apple," *Guardian* (London), February 1, 2002.

Gulf Power . . . pled guilty: Michelle Hiskey, "Artist Who Fled Grand Jury Pleads Guilty on Tax Charge," *Atlanta Journal-Constitution,* April 23, 1990.

a discrimination lawsuit: Russell Grantham, "Four Join Suit Alleging Bias by Southern," *Atlanta Journal-Constitution,* August 15, 2000.

In the 1990s: Dan Rather, "Georgia Power Hears Discrimination Charges," *CBS Evening News with Dan Rather,* February 25, 2002.

157 *"no earthly idea":* Bill Dahlberg, quoted in Bob Edwards, "Georgia Power Hears Discrimination Charges," *Morning Edition,* National Public Radio, April 30, 2001.

"When children can't breathe": Janet Reno, quoted in Bruce Barcott, "Changing All the Rules," *New York Times Magazine,* April 4, 2004.

159 *the company had to build:* Eliot Spitzer, attorney general of the State of New York, testimony before Senate Committee on Environment and Public Works and Senate Committee on the Judiciary, July 16, 2002.

pollution from power plants: Abt Associates, "Particulate-Related Health Impacts of Eight Electric Utility Systems" (report, prepared for Rockefeller Family Funds, April 2002).

160 *"This was the most":* Sylvia Lowrance, quoted in Barcott, "Changing All the Rules."

"It was clear": Interview with Eric Schaeffer, 2002.

According to one study: U.S. Public Interest Research Group, "Abuse of Power: Southern Company's Campaign to Undo, Weaken, Delay and Circumvent Life-Saving Pollution Rules," May 2001.

161 *between 1970 and 1990:* U.S. Environmental Protection Agency, "The Benefits and Costs of the Clean Air Act, 1970–1990" (report, prepared for U.S. Congress, October 1997).

"A Citizen Wherever We Serve": This slogan is attributed to Preston Arkwright, the first president of Georgia Power. It is widely used at Southern as the "guiding principle" for how the company conducts business. See David Ratcliffe, "Our Energy Future: A Delicate Balance" (speech, Rotary Club of Atlanta, June 13, 2005).

162 *"The suit basically alleges"*: Interview with Bruce Buckheit, 2003.

On March 25: Gertrude Trawick, "Area Power Facility to Cost $1 Billion," *Macon (Ga.) News,* March 26, 1974.

According to an outside audit: O'Brien-Kreitzberg and Associates, "Retrospective Audit Report for Plant Scherer Units 3 and 4" (report, Georgia Public Service Commission, April 6, 1987), pp. 4-7, 4-17.

163 *roughly five times as dirty:* Accurate sulfur dioxide emissions data for Plant Scherer in 1989 is not publicly available. For this calculation, I used 1996 data, which is likely more favorable to Georgia Power than the earlier data. According to the EPA's eGRID database (www.epa.gov/cleanenergy/egrid/index.htm), Plant Scherer's 1996 sulfur dioxide emission rate was .90 pounds per million BTU. Determining EPA's Best Available Control Technology standard for 1989 is more complex, in part because BACT standards vary according to coal type and power-plant design. The best estimate I'm aware of can be found in the April 2002 testimony of Matt Haber, a staff engineer at the EPA, in the case *United States* v. *Illinois Power Company and Dynegy Midwest Generation, Inc.,* in the U.S. District Court for the Southern District of Illinois. During the trial, which concerned New Source Review violations at the Baldwin power plant in Baldwin, Illinois, Haber testified that the 1988 BACT standard for sulfur dioxide emissions from a 585-megawatt coal plant — slightly smaller than each unit at Plant Scherer — was .30 pounds per million BTU. The 1989 standard (about which Haber did not testify) would have undoubtedly been lower.

a spokesperson: Lolita Browning Jackson, e-mail communication with the author, January 5, 2006.

164 *On November 4:* O'Brien-Kreitzberg, "Retrospective," pp. 4-10, 4-17.

"work was resumed": Georgia Power 1976 Annual Report, pp. 12–13.

In 1979, Georgia Power's: O'Brien-Kreitzberg, "Retrospective," p. 5-2.

"The commercial operation": Ibid., p. 1-3.

at least twenty-two power plants: Kennedy and Associates, "Plant Scherer Coal Supply Investigation" (report, Georgia Public Service Commission, July 1987), p. II-6.

165 *A Gallup poll:* Barcott, "Changing All the Rules."

"If there's any": George W. Bush, quoted in ibid.

A March 23: Jena Heath, "Southern's Fingerprints Seen on Energy Policy," *Atlanta Journal-Constitution,* March 28, 2002.

166 *Between 1998 and 2004:* Center for Public Integrity, *LobbyWatch,* www.publicintegrity.org/lobby/profile.aspx?act=clients&year=2003&cl=L002762.

"did not dwell": Scott Segal, quoted in Christopher Drew and Richard Oppel, "How Industry Won the Battle of Pollution Control at EPA," *New York Times,* March 6, 2004.

"As we discussed": Christine Todd Whitman, quoted in ibid.

167 *"emits as much"*: U.S. Public Interest Research Group, "Abuse of Power."

"Inside the company": Interview with former Southern Company executive, 2004.

168 *five times the rate*: Calculated from the U.S. Environmental Protection Agency's Clean Air Markets database, cfpub.epa.giv/gdm, December 20, 2005. To be fair to the Southern Company, this calculation was made from the plant's sulfur dioxide emissions rate, not total tons emitted. (Because Scherer is one of the largest coal plants in the nation, comparisons of total tonnage of pollutants sometimes make the plant look dirtier than it is.) According to *Dirty Kilowatts: America's Most Polluting Power Plants,* a report published in May 2005 by the Environmental Integrity Project in Washington, D.C., www.environmentalintegrity.org/pub314 .cfm, that ranked the pollution rates of the 359 largest power plants in the United States, Scherer rates as the twenty-eighth dirtiest power plant in the nation if measured by the total tons of sulfur dioxide released in 2004 and the one hundred sixty-third dirtiest if measured by the amount of sulfur dioxide released per megawatt-hour of operation. Several of Southern Company's other coal plants score much higher on the list, including Plant Gaston near Wilsonville, Alabama, which is ranked ninth dirtiest by number of tons released and twenty-eighth dirtiest per megawatt-hour of operation, and Plant Bowen near Cartersville, Georgia, which ranked second in tonnage and fifty-third per megawatt-hour. Both Gaston and Bowen are among the coal plants cited in the U.S. Department of Justice's New Source Review lawsuit against the Southern Company.

In terms of other pollutants, Plant Scherer does not rank in the top fifty on the *Dirty Kilowatts* list either in tonnage released or emissions rate for nitrogen oxides. With mercury, Scherer rates tenth in total pounds released and one hundred sixth in pounds released per megawatt-hour. With carbon dioxide, Scherer is one hundred twenty-eighth in pounds emitted per megawatt-hour and number one in the nation — releasing an astounding 25.5 million tons in 2004 — in annual carbon dioxide emissions.

the Atlanta Journal-Constitution: Margaret Newkirk, "Southern Co. Hires Clean Air Act Regulator," *Atlanta Journal-Constitution,* September 4, 2003.

169 *"We of course"*: Charles Peters, "Southern Airs," *Washington Monthly,* October 2003.

8. Reversal of Fortune

173 *Several hurricane experts:* Kerry Emmanuel, "Increasing Destructiveness of Tropical Cyclones Over the Past Thirty Years," *Nature* 436, July 2005,

pp. 686–88; P. J. Webster et al., "Changes in Tropical Cyclone Number, Duration, and Intensity in a Warming Environment," *Science*, September 16, 2005, pp. 1844–46. See also Roger Pielke Jr., "Are There Trends in Hurricane Destruction?" *Nature* 438, December 2005, p. E11.

174 *"This is one"*: Interview with Lloyd Keigwin, June 2004.

175 *"We're walking around"*: Interview with James Rogers, July 2001.

I interviewed Rogers: Jeff Goodell, "How Coal Got Its Glow Back," *New York Times Magazine*, July 22, 2001.

"one of the most": Paul Anderson, "Taking Responsibility" (speech, Charlotte Business Journal's Tenth Annual Power Breakfast, April 7, 2005).

"Give us a date": Wayne Brunetti, quoted in John Carey, "Global Warming," *BusinessWeek*, August 16, 2004.

"But I have a family": Personal communication.

177 *melting permafrost*: Fred Pearce, "Climate Warning as Siberia Melts," *New Scientist*, August 11, 2005, p. 12.

the Swedish scientist: Spencer Waert, *The Discovery of Global Warming* (Boston: Harvard University Press, 2003). Available online (June 2005 version) at www.aip.org/history/climate/index.html.

the level of CO_2: National Oceanic and Atmospheric Administration, Global Monitoring Division, "After Two Large Annual Gains, Rate of CO_2 Increase Returns to Average," March 31, 2005.

the average temperature: Intergovernmental Panel on Climate Change, *Climate Change 2001: Synthesis Report*, September 2001, p. 8.

By 2100, unless something: Ibid.

"The greenhouse effect": James Hansen, quoted in Ira Boudway et al., "Climate Warriors and Heroes," *Rolling Stone*, November 17, 2005.

178 *"highly adverse consequences"*: National Academy of Sciences, Geophysics Research Board, *Energy and Climate: Studies in Geophysics* (Washington, D.C.: National Academy of Sciences, 1977).

"I've never seen": Michael Oppenheimer, quoted in John Noble Wilford, "His Bold Statement Transforms the Debate on Greenhouse Effect," *New York Times*, August 23, 1988.

179 *"Those who think"*: "Global Lukewarming," Editorial, *New York Times*, November 5, 1989.

"the environmental president": Philip Shabecoff, "In Thicket of Environmental Policy, Bush Uses Balance as His Compass" *New York Times*, July 1, 1990.

"disinterest and opposition": Quoted in Donald A. Brown, *American Heat, Ethical Problems with the United States' Response to Global Warming* (Lanham, Md.: Rowman and Littlefield, 2002), p. 21.

Sununu's political opposition: William Nitze, "A Failure of United States Leadership," in *Negotiating Climate Change*, ed. Irving M. Mitzner and J. Amber Leonard (Cambridge: Cambridge University Press, 1994), pp. 187–200.

180 *"Stabilization of greenhouse gas"*: United Nations Framework Convention on Climate Change, 1992, article 2, p. 9.

"You need to take": Harlan Watson, quoted in Jeremy Leggett, *The Carbon War* (New York: Penguin Books, 1999), pp. 128–29.

"It lulled people": Ross Gelbspan, *The Heat Is On* (New York: Perseus Books, 1998), p. 9.

"There has been": Quoted in ibid., p. 36.

181 *"reposition global warming"*: Ibid., p. 34.

Western Fuels also spent: Ibid., p. 36.

"The Western Fuels campaign": Ibid.

182 *"It is easy"*: Fred Palmer, "Fossil Fuels or the Rio Treaty — Competing Visions for the Future" (speech, Coaltrans 96, Barcelona, October 21, 1996).

credible estimates: Jason F. Shogren, "Kyoto Protocol: Past, Present, and Future," *AAPG Bulletin* 88, no. 9 (September 2004): 1221–26.

Ohio coal operator: Robert E. Murray, testimony before House Committee on Government Reform and Oversight, Subcommittee on National Economic Growth, Natural Resources, and Regulatory Affairs, June 24, 1998.

183 *$1.8 trillion:* Resource Data International, "Putting U.S. Electricity Supply and GDP at Risk," report presented to Peabody Holding Company, February 1998, www.peabodyenergy.com/pdfs/Kyoto.pdf.

Kyoto would increase: Peabody Group, "Buenos Aires Update: Clearing the Air About the Greenhouse Effect," Peabody Group information sheet, February 1998, www.peabodyenergy.com/pdfs/BuenoAir.pdf.

184 *"There was clearly a moral edge"*: Personal communication.

"The 2000 election": Interview with Jack Gerard, July 2001.

185 *"Global warming"*: "The 2000 Campaign; Second Presidential Debate Between Gov. Bush and Vice President Gore," *New York Times,* October 12, 2000.

"Vice President Gore": Quoted in Andrew Revkin, "Despite Opposition in Party, Bush to Seek Emissions Cuts," *New York Times,* March 10, 2001.

186 *"We knew Bush"*: Personal communication.

"to enter a new era": Christine Todd Whitman, statement before the U.S. Senate Committee on Environment and Public Works, Washington, D.C., January 17, 2001, epw.senate.gov/107th/whi_0117.htm.

"particularly well-represented": John Mintz, "Transition Advisers Have Much to Gain," *Washington Post,* January 17, 2001.

187 *in the 2000 election:* Center for Responsive Politics, "Coal Mining: Top Contributors to Federal Candidates and Parties," 2000, www.opensecrets .org/industries/contrib.asp?Ind=E1210&Cycle=2000.

"an all-American story": Irl Engelhardt, quoted in Kevin Kipp, "High Energy," *St. Louis Commerce Magazine,* March 2002.

His father was killed: Linda Tucci, "Coal's Idealistic Side Keeps Peabody Chief Passionate," *St. Louis Post-Dispatch,* May 2, 2003.

188 *Exactly what was discussed:* Bill Mears, "High Court Hears Arguments on Cheney Task Force," June 24, 2004, www.cnn.com/2004/LAW/04/27/scouts.cheney/index.html.

189 *as Dr. Fred Singer:* Fred Singer (speech, Electric Power, Sixth Annual Conference and Exhibition, March 30–April 1, 2004, Baltimore).

190 *Wally Broecker hypothesized:* Wallace Broecker, "The Great Ocean Conveyor," *Oceanography* 4, no. 2 (1991): 79–89. See also Wallace Broecker, "Thermohaline Circulation, the Achilles Heel of Our Climate System," *Science* 278 (1997): 1582–88.

191 *Richard Alley determined:* Richard Alley, *The Two-Mile Time Machine: Ice Cores, Abrupt Climate Change, and Our Future* (Princeton, N.J.: Princeton University Press, 2000), p. 4.
"You might think": Ibid., p. 83.

192 *"I was keenly aware":* Christine Todd Whitman, *It's My Party Too* (New York: Penguin Press, 2005), p. 169.
Bush's mandatory cap: Ibid., p. 170.
"George Bush was": Christine Todd Whitman, quoted in Ron Suskind, *The Price of Loyalty* (New York: Simon and Schuster, 2004), p. 102.
"There's no question": Ibid., p. 99.

193 *"The climate is changing":* Christine Todd Whitman, quoted in Dick Thompson, "Will Bush Turn Green?" *Time,* March 12, 2001.
"a PR tool": Personal communication.
"it doesn't take": Thomas Kuhn, quoted in Bill Dawson, "The Politics of Energy: Coal," report, Center for Public Integrity, December 3, 2003.
"This is a colossal mistake": Myron Ebell, quoted in Andrew Revkin, "Despite Opposition."

194 *On March 1:* Drew and Oppel, "How Industry Won the Battle."
"technological genocide": Arthur Robinson and Jane Orient, "Science, Politics, and Death," *New American,* June 14, 2004.

195 *"there is no convincing":* The petition is posted on Arthur Robinson's Web site, www.oism.org/pproject/s33p37.htm.
"the theology of global warming": James Schlesinger, "The Theology of Global Warming," *Wall Street Journal,* August 8, 2005.
"extraordinary response": Chuck Hagel, quoted in "Oregon Institute of Science and Medicine," *SourceWatch,* www.sourcewatch.org/index .php? title=Oregon_Institute_of_Science_and_Medicine.
Scientific American *took:* George Musser, "Climate of Uncertainty," *Scientific American,* October 2001.
"but it's really just me": Interview with Arthur Robinson, September 2005.

196 *$575,000 settlement:* Arthur Robinson, "100 Man Years of Data," *Access to Energy,* June 1996.

"a hypothesis": Robinson, interview.

the views he expressed: Robinson and Orient, "Science, Politics and Death."

197 *"I'm sorry to have":* Robinson, interview.

The Financial Times *noted:* Suskind, *The Price of Loyalty,* p. 109.

198 *"I would strongly recommend":* Whitman, *It's My Party,* p. 173.

"clarification of": Quoted in Suskind, *The Price of Loyalty,* pp. 118–19.

199 *"You know what?":* Quoted in ibid., p. 120.

"Apparently, everyone": Whitman, *It's My Party,* p. 175.

"Christie, I've already": George W. Bush, quoted in Suskind, *The Price of Loyalty,* p. 121.

200 *a study by:* David Hawkins, "Harmonizing the Clean Air Act with Our Nation's Energy Policy," testimony before Senate Committee on Environment and Public Works, Subcommittee on Clean Air, Wetlands, Private Property, and Nuclear Safety, March 21, 2001.

"wind dummy": Colin Powell, quoted in Mike Allen, "Cabinet Chess Begins Anew," *Washington Post,* November 22, 2002.

201 *The company offered:* "Peabody Energy Stock Soars at Initial Public Offering," *New York Times,* May 23, 2001.

as of May 31: Securities and Exchange Commission, Peabody Energy form 10-K/A (2001), item 12, "Security Ownership of Certain Beneficial Owners and Management."

By the end of 2005: This calculation was made from records of insider stock transactions provided by EDGAR Online, www.finance.yahoo.com, October 19, 2005, and December 15, 2005. The total amount includes revenues from the sales of Engelhardt's direct and indirect holdings (indirect holdings include stock that is held in a trust or limited partnership). Also a portion of this $55 million total may reflect gains on stock that Engelhardt purchased at an earlier date for investment purposes. To be sure that it was correct to count both direct and indirect holdings in my tally of Engelhardt's stock sales, I called the Securities and Exchange Commission; the public relations officer suggested I consult Alan Dye, a noted federal securities lawyer in Washington, D.C. Dye explained that indirect holdings are frequently used by executives for a variety of legal reasons, including avoiding estate taxes. A person benefits economically from both direct and indirect holdings, Dye said, adding that both are fairly considered part of a CEO's executive compensation.

9. The Coal Rush

203 *2,500 construction jobs:* Jim Muir, "Jobs Coming to Region: Governor Visits Nashville to Tout Prairie State Energy Campus," *Southern Illinoisan* (Carbondale, Ill.), February 8, 2005.

more than twice as much: Environmental Defense, *Automakers' Corpo-*

rate Carbon Burdens, 2005. According to this report, 5 million tons of CO_2 was emitted by vehicles sold by Ford in 2003.

204 *"the testicular virility":* Rod Blagojevich, quoted in Maureen O'Donnell, "'Testicular Virility' Lets Gov Face Mell," *Chicago Sun-Times,* May 17, 2005.

205 *fourteen hundred 1,000-megawatt:* Connie Holmes, National Mining Association, cited in "Coal — A Twenty-First-Century Fuel," congressional briefing, May 19–20, 2005.
China is investing: "China Considers 24 Billion USD Investment in Coal-to-Liquids Technology," *AFX News,* September 27, 2005.

206 *"This is not":* Interview with Bret Clayton, June 2004.
"To put it plainly": Interview with Tom Burke, September 2005.
the world as we know it: "Report of the Steering Committee," Avoiding Dangerous Climate Change, International Symposium on the Stabilization of Greenhouse Gas Concentrations, Exeter, England, February 1–3, 2005.

207 *"Climate change is real":* Rajendra Pachauri, quoted in Geoffrey Lean, "Global Warming Approaching Point of No Return, Warns Leading Climate Expert," *Independent* (London), January 23, 2005.
to have a likely chance: Malte Meinshausen, "The Risk of Overshooting 2° C" (presentation, Avoiding Dangerous Climate Change, February 2, 2005).

209 *"Most power plants":* Interview with Bill Hoback, July 2005.
"The only question": Interview with Michael Skelly, October 2005.

211 *"the future of coal":* Quoted in Kenneth J. Stier, "Dirty Secret: Coal Plants Could Be Much Cleaner," *New York Times,* May 22, 2005. See also Robert Williams, "IGCC: Next Step on the Path to Gasification-Based Energy from Coal."

212 *"While we support":* Irl Engelhardt, quoted in "Global Warming: Can We Find Common Ground?" *Cinergy 2004 Annual Report,* February 2005, www.cinergy.com/pdfs/reports/04sar/suppliers.pdf.

Starting in June 2005, I made a number of requests to speak directly with Peabody executives about the company's decision to build a conventional coal plant instead of an IGCC plant in Illinois. I got nowhere. On November 15, 2005, I submitted ten written questions to the company's spokesperson, Vic Svec; I received a response on December 21, 2005. The replies to my questions were brief and, in most cases, quite vague. Svec asserted that "advanced combustion plants like Prairie State represent leaps in technologies" that are "moving us toward the ultimate goal of near-zero emissions for electricity from coal." He added that Peabody advocates a "balanced approach toward carbon management" that will improve the scientific understanding of climate change, advance technologies to capture and sequester CO_2, and promote energy efficiency. The e-mail was followed up by a phone call from Svec, but, like the e-mail replies, our conversation never moved much beyond general assertions and boilerplate PR rhetoric. More information about Peabody's views can be found at www.peabody energy.com.

213 *"The coal industry":* Interview with Dale Simbeck, July 2005.

214 *"It's AEP's opinion":* Bruce H. Braine and Michael J. Mudd, "Integrated Gasification Combined Cycle," American Electric Power Corporation White Paper, May 5, 2005.

215 *The Energy Policy Act:* Taxpayers for Common Sense, "Authorized Spending in the Energy Bill Conference Report" July 28, 2005, www.taxpayer .net/energy/pdf/hr6finalanalysis.pdf.

216 *The plant recently:* "Coal Gasification: Ready for Prime Time," *Power,* March 2004.

"Coal is": Interview with Robert Ukeiley, May 2005.

217 *"It's hard to fight":* Interview with Jesse Ausubel, November 2004.

"Hawkins's real agenda": Interview with Vic Svec, December 2005.

218 *"When people":* Burke, interview.

billion dollars: Elizabeth Souder, "Legislators Slam Record TXU profits," *Dallas Morning News,* November 10, 2006.

219 *More than 90 percent:* U.S. Department of Energy, Office of Fossil Energy, "The Early Days of Coal Research," www.fe.doe.gov/aboutus/history/syntheticfuels_history.html.

special tax breaks: Donald L. Bartlett and James B. Steele, "The Great Energy Scam," *Time,* October 4, 2003.

"The entire synfuels industry": Chuck Lindell, "Doggett Aims to Kill Energy Tax Break," *Austin American-Statesman* (Tex.), April 28, 2004.

220 *Schweitzer has proposed:* John S. Adams, "Coal Rush," *Missoula* (Mont.) *Independent,* September 29, 2005.

"Like all Americans": Brian Schweitzer, "The Other Black Gold," *New York Times,* October 3, 2005.

221 *three and a half barrels of water:* Interview with Andre Steynberg, November 2005. According to Steynberg, an 80,000-barrel-per-day plant typically consumes 1,200 to 1,800 cubic meters of water per hour. Converting cubic meters to gallons and then to barrels, that's 271,992 barrels of water a day. This consumption can be reduced somewhat by using air cooling (instead of water) at the plant, but Steynberg says that that would increase costs and lower the overall efficiency of the plant, which would increase coal consumption and raise carbon dioxide emissions.

the amount of CO_2: Robert H. Williams and Eric D. Larson, "A Comparison of Direct and Indirect Liquefaction Technologies for Making Fluid Fuels from Coal," *Energy for Sustainable Development,* December 2003.

using typical grid power: Joseph Romm, cited in James Cascio, "Gas Optional & Green," WorldChanging.com, June 14, 2005, www.world changing.com/archives/002891.html.

222 *"The whole notion":* Robert H. Williams, "IGCC: Next Step on the Path to Gasification-Based Energy from Coal," report, National Commission on Energy Policy, November 2004. See also Emma Young, "Going Underground," *New Scientist,* September 3, 2005, p. 34.

223 *production in the oil field:* John K. Borchardt, "A Greenhouse Gas Goes Underground," *Christian Science Monitor,* October 28, 2004.

there are more than enough: James Dooley, "Regional Assessments of the Geologic CO_2 Sequestration Resource" (presentation, Third U.S.-China Clean Energy Workshop, Morgantown, W.Va., October 18–19, 2004). See also James Dooley, "A First-Order Global Geological CO_2-Storage Potential Supply Curve and Its Application in a Global Integrated Assessment Model" (paper, Seventh International Conference on Greenhouse Gas Control Technologies, Vancouver, Canada, September 5, 2004).

224 *raise the price of electricity:* Ram Narula, "IGCC vs. SCPC: Battle of Technologies II" (presentation, Gasification Technologies Conference, San Francisco, Calif., October 2005, www.gasification.org). The cost of capturing and storing carbon is the subject of much debate and analysis, in part because the technology is so new, and in part because cost estimates for carbon storage vary from region to region. As a general guideline, however, Narula, who is the manager of technology for Bechtel Power Corporation, estimates the cost of electricity from a new supercritical combustion coal plant to be 4.1 cents per kilowatt-hour without carbon capture and storage, and 6.9 cents with it; from an IGCC plant, he estimates 4.4 cents per kilowatt-hour without capture and storage, and 6 cents with it. Narula's analysis is in line with other studies, which conclude that it will cost about twice as much to capture carbon from a combustion coal plant as it will from an IGCC plant. The cost of carbon storage will then depend on the geology of the area where the plant is located, and on whether the CO_2 can be used to enhance oil recovery in nearby oil fields. See also "Carbon Capture and Storage," Working Group III of the Intergovernmental Panel on Climate Change, p. 347, table 8.3a.

infrastructure required: Interview with Lynn Orr, October 2006.

"Any time you inject": Interview with Bill Evans, November 2005.

225 *At Mammoth:* Kevin Coughlin, "Death of Skier Points to Invisible Danger," *Newark Star-Ledger,* March 6, 2005.

10. The Frontier

226 *Topsoil is blowing:* China News Service, "Expert Says That Xinjiang Is a Major Sandstorm Source Area of China," March 20, 2001; Lester Brown, "China Is Losing the War on Advancing Deserts," *International Herald Tribune,* August 13, 2003.

glaciers are melting: Xinhua News Agency, "Glaciers Fading Away in Xinjiang," November 9, 2004.

229 *Outdoor air pollution: Comparative Quantification of Health Risks,* p. 1402, table 17.8.

Acid rain is falling: "The Chinese Miracle Will End Soon" (interview with Pan Yue, deputy director of China's State Environmental Protection Administration), *Der Spiegel,* March 7, 2005.

fifteen thousand people: Howard French, "Riots in Shanghai Suburb as Pollution Protest Heats Up," *New York Times,* July 18, 2005.

in the nearby city of Dongyang: Clifford Coonan, "Chinese Farmers Riot over Crop Poisoning," *The Times* (London), April 12, 2005.

"We are convinced": "The Chinese Miracle."

"Chinese leaders": Interview with Husayn Anwar, January 2003.

230 *The coal-fired power plants:* Mark Clayton, "New Coal Plants Bury Kyoto," *Christian Science Monitor,* December 23, 2004.

China has already: Thomas Friedman, "China's Little Green Book," *New York Times,* November 2, 2005; Howard French, "In Search of a New Energy Source, China Rides the Wind," *New York Times,* July 26, 2005.

231 *In July 2005, the administration:* Jane Perlez, "U.S. to Join China and India in Climate Pact," *New York Times,* July 27, 2005.

"The [Asia-Pacific] pact": Senator John McCain, quoted in Amanda Griscom Little, "Pact or Fiction," *Grist,* August 4, 2005, http://www.grist.org/news/much/2005/08/o4/little-pact/.

233 *"FutureGen road show":* Personal communication.

"wreck our economy": "The Bush Interview: Tonight with Trevor McDonald," *Guardian* (London), July 4, 2005.

"I hear about": Interview with Douglas Ogden, September 2005.

"What [developed nations]": Quoted in Shogren, "Kyoto Protocol," p. 1223.

236 *"the greatest green":* "The Invisible Green Hand," *Economist,* July 4, 2002.

as much as $12.5 billion: "In Asia, a Hot Market for Carbon," *BusinessWeek,* December 12, 2005.

239 *10 percent of the forest:* Judith Shapiro, *Mao's War Against Nature* (Cambridge: Cambridge University Press, 2001), p. 82.

240 *"a form of brain damage":* David Brower, "It's Healing Time on Earth" (Twelfth Annual E. F. Schumacher Lecture, Stockbridge, Mass., October 1992).

"a paradigm shift": Interview with Eileen Claussen, March 2003.

In New Zealand: W. Wayt Gibbs, "How Should We Set Priorities?" *Scientific American,* September 2002.

241 *Between 1992 and 2004:* Daphne Wysham, "Emissions Creep," *Grist,* March 25, 2005.

242 *"I think the free hand":* Interview with Dai Qing, January 2003.

244 *negotiators fought hard:* See Donald A. Brown, *American Heat,* pp. 13–42.

245 *one third of the nation's farmland:* Robert B. Jackson and William H. Schlesinger, "Curbing the U.S. Carbon Deficit," *Proceedings of the National Academy of Sciences,* November 9, 2004, pp. 15827–29.

"Right now, these rain-forest nations": Joseph Stiglitz, quoted in Miguel Bustillo, "Rain Forest Nations Seek Incentive to Conserve," *Los Angeles Times,* November 27, 2005.

Epilogue

250 *"All left wing":* George Orwell, "Rudyard Kipling," in *Critical Essays* (London: Secker and Warburg, 1947).

253 *"That debate's history":* Alex Steffen, "New Orleans: Everything Has Changed," WorldChanging.com, September 5, 2005.

254 *"The arguments against slavery":* Ian Frazier, quoted in Dave Weich, "Ian Frazier's Heroes," Powells.com, February 17, 2000, http://www.powells.com/authors/frazier.html.

Afterword

257 *a billion tons:* "Coal usage expected to swing up again in EIA's two-year forecast," *Platt's Coal Trader,* January 10, 2007.

"different planet": quoted in "If we fail to act, we will end up with a different planet," *The Independent* (U.K.), January 1, 2007.

258 *In 2000, CO_2 emissions:* Richard Black, "Carbon Emissions Show Sharp Rise," BBC News, November 27, 2006.

rise by one third: Energy Information Administration, *Annual Energy Outlook 2007 (Early Release),* December 2006.

Arctic sea ice: Randolph Schmid, "Signs of Warming Continue in the Arctic," *Associated Press,* November 17, 2006.

259 *ocean acidification:* James C. Orr, et al, "Anthropogenic Ocean Acidification over the Twenty-First Century and Its Impact on Calcifying Organisms," *Nature,* vol. 437 (2005).

"Stern's numbers": David Roberts, "We Got Our Kicks in 2006," *Grist,* December 22, 2006. http://www.grist.org/news/maindish/2006/12/22/top10/.

"You have to be crazy": Interview with Vinod Khosla, December 2006.

260 *"The science of global warming is clear":* "CEOs Plead for Mandatory Emissions Caps," *New York Times,* January 22, 2007.

Index